COSMIC VISIONS WITHIN THE MICROCOSM OF MY RIGHT HEMISPHERE: A new theory on the functions of black holes and the development of the cosmic brain

VINCENT L. DI PAOLO

authorHOUSE

AuthorHouse™
1663 Liberty Drive
Bloomington, IN 47403
www.authorhouse.com
Phone: 1 (800) 839-8640

© *2018 Vincent L. Di Paolo. All rights reserved.*

No part of this book may be reproduced, stored in a retrieval system, or transmitted by any means without the written permission of the author.

Published by AuthorHouse 09/25/2018

ISBN: 978-1-5462-6119-3 (sc)
ISBN: 978-1-5462-6120-9 (hc)
ISBN: 978-1-5462-6118-6 (e)

Library of Congress Control Number: 2018911275

Print information available on the last page.

Any people depicted in stock imagery provided by Getty Images are models, and such images are being used for illustrative purposes only. Certain stock imagery © Getty Images.

This book is printed on acid-free paper.

Because of the dynamic nature of the Internet, any web addresses or links contained in this book may have changed since publication and may no longer be valid. The views expressed in this work are solely those of the author and do not necessarily reflect the views of the publisher, and the publisher hereby disclaims any responsibility for them.

CONTENTS

Dedication vii
Special Dedication ix
Preface xi

Chapter 1	Creation: the First Big Bang 1	
Chapter 2	Gravity: How it rules the cosmos and our very lives 11	
Chapter 3	Schemata of a Galaxy and the Function of its Galactic Black Hole 16	
Chapter 4	Black Holes: Super machines programmed from the beginning of time to recycle stars, galaxies, and universes. 19	
Chapter 5	Probability of Sun-like Stars in our Universe. 29	
Chapter 6	Early Development of Galactic Black Holes 32	
Chapter 7	Earth: a Beautiful Living Planet 36	
Chapter 8	The Human Brain: a cosmic neurological development and a microscopic reflection of the universe 44	
Chapter 9	Myelin Permits Higher Intelligence 53	
Chapter 10	Myelinations I & II: Insulation of the Left Hemisphere . 56	
Chapter 11	Myelinations III & IV: Insulation of the Right Hemisphere. 68	
Chapter 12	Nurturing your Brain for Life. 71	

Chapter 13 Inherited Cosmic Intelligence 73
Chapter 14 Expanding Cosmos, Expanding Cosmic Brain 75

Epilogue 77
Glossary 83
Bibliography 105
Credits for Photographs, Paintings, Drawings and Charts 121
About the Author 123

DEDICATION

I dedicate this book to the memory of my father, Lorenzo, who understood my visions, to my mother, Maria Giovanna, and to my children, Alexandra and David, who suffered much time listening to my theoretical visions of the cosmos and the development of the human (cosmic) brain.

SPECIAL DEDICATION

I also dedicate this book to the memory of Stephen Hawking, who shared so many of my visions of the cosmos and the functions of black holes. He was a great man with a brilliant mind and vision, who conquered his amyotrophic lateral sclerosis by continuing to teach and write many wonderful books in the past forty years. His visions paralleled many of mine; and I hope that I have answered some of his questions, especially "What happens to the information of those star systems as they are taken in to be recycled by galactic black holes?" The memories of his brilliant mind are now mixed with mine, deep in the libraries of my mind.

Vincent L. Di Paolo (2018)

PREFACE

What I am about to tell you is the very first time that I reveal my incredible dreams of my youth, which became my visions as an adult. The first memory that I can recall of the dreams that occupied much of my sleep was back in 1951 at the tender age of two. At first, I saw those dreams as beautiful fireworks in the dark night sky. With time, those dreams became clearer: the objects flying in the dark skies were moons, planets and stars. As a five year old child I prayed to God to allow me to continue dreaming about the universe, which was the best word I then knew to describe the ever expanding cosmos.

I was born in 1949, in Cansano, a small tenth century medieval town in the province of Aquila, built on the ruins of the ancient village of Ocriticum on a plateau of the Central Apennines. The plateau is surrounded by beautiful snowcapped mountains all year round. The ruins in Cansano date more than 2000 years before Rome. As a child, my main preoccupations, after school, were exploring ruins and fine arts. As a direct descendant of Giovanni Di Paolo, I began drawing as early as two years old. From 1955 to 1959 I was a young student in an Italian school based on Jean Piaget's theories of education. By second grade we knew the four mathematical operations by heart; and in third grade we applied the four operations to plane geometry. By fifth grade we were learning pre-algebra, world history and geography, sciences, Latin, creative writing, music and fine arts, and of course, soccer. By then, I was hooked on mathematics, geometry, the sciences, writing and fine arts. In 1955 I took first place in a national art contest for elementary school students; and, in 1959, I won first place in another national contest once again.

By the age of eight, my dreams had become so realistic and frightening that many times I woke up sweating and so scared that I would possibly die. Also, a new sensation had become recurrent in each and every dream: I was able to fly. The first time I experienced flying in my dreams I was barely four years old; and it was the origin of a progressively exciting sensation that developed into cosmic flying by the time I was ten years old. I can clearly recall my first flight from the roof of my house, gliding over the cobble stone street which led into the town piazza, where I would land on my feet. By five years old I would climb the hills that surrounded my town and I would fly down, soaring over the plateau below as my dreams became more daring. By seven years old I would climb Mount Everest in my vivid dreams and I would dive, fly over the oceans, landing in other continents. In some of my recurring dreams I would get to the top of the Empire State Building, climb to the very top of the antenna, where I could barely stand with my feet together; from the top of the antenna I would dive, flying low over the long busy streets of New York, flying faster than the speed of sound over the Atlantic Ocean and land in the middle of the piazza in Cansano, Italy.

By the time I was ten years old my dreams had become so exciting that I would wake up and feel my heart beating so fast as if I had just finished running a marathon race. By then, in my dreams, I was flying to the moon, around the moon and landing on my two feet on its dusty soil, where I would skip along fifty feet per step; and, when I was ready to come back to earth I would jump on both feet and push into space, flying faster than a rocket landing on the piazza of my town. Space flying in my dreams became more daring as I flew to the Asteroids, where I would asteroid hop around the asteroid belt. I felt like the Little Prince in Antoine De Saint-Exupéry's beautiful and deeply meaningful story. Along the way I would make a stop on Mars, where I dreamed of climbing Olympus Mons and dive into the Martian landscape and back to earth, gently landing on my town's main piazza. My cosmic dreams became more vivid and more daring where I would fly at the speed of light, disintegrate into light, reaching distant star systems of the Milky Way, and reintegrate into my body as I approached the earth. There wasn't a night in my youth that I did not dream cosmic traveling, reaching the farthest star system of our galaxy and flying to neighboring galaxies.

By ten years old, I felt that I really didn't belong to planet earth; I imagined that I was from a distant planet in a star system of a galaxy far away from the Milky Way. To this very day, I still wonder where I might have come from, still searching for a meaning of my very existence in this vast cosmos that seems to have no end. As an adult, those daily dreams of my youth rarely reoccurred;

however, they are forever molded in my memory and easily recalled in vivid visions within the extremely active right hemisphere of my brain.

As a young student in Cansano, Italy, I was very studious, absorbing everything that my teachers taught me and reading about things that interested me. As children we enjoyed exploring around the ancient parts of Cansano, hoping of discovering treasures from stories our grandparents had retold us many times. While exploring the countryside in 1956, two friends and I had found a hole in a field in the Pantano, a small prairie at the foothills of Colle Mitra. We went home to get candles and matches and we immediately went back to the hole which we daringly entered. We lit the candles which gave us some light to see where the opening would take us. As we descended down the hole it was clear that we were in a wide corridor with stone walls on each side: we were standing on a subterranean narrow street of the ancient Ocriticum. On several occasions we explored on, finding a few coins, broken pottery and bones. The following summer of 1957 we went back but we could not find the hole: it had been covered with soil and wild grasses camouflaged the exact spot on the long grassy field which followed the foothills of the mountains overlooking the valley of Sulmona. It is now the archaeological site of Ocriticum.

In the spring of 1959, having said good-bye to all my dear friends and relatives, my family and I boarded an ocean liner in Naples, which would carry us to Venezuela where my father was working. Our first stops were the ports of Barcelona and of Santa Cruz in Tenerife, the capital of the Canary Islands off the west coast of Morocco. There I first tasted delicious kumquats, sold in little baskets by fruit vendors. Also, alongside our ship, young boys dove from their small dinghies for silver coins. I threw three of my silver coins, one at the time, into the waters just to see how far those boys, my age, would dive down, catch the coin and come back up smiling holding the coin with their fingers. It was thrilling and fascinating for me to see how those brave young boys would dive dozens of times so deep for silver coins, which most probably helped their families survive. That night, I dreamed that I repeatedly dove off from the top of the bulwarks into the waters for silver dollars and I would climb the rope that had been placed for me by one of the officers. I really enjoyed dining in the great dining hall, where I would dine with my family and once again with the family of my new friends I had made on board. Finally, after seven days of sailing across the Atlantic Ocean we arrived at La Guaira, the seaport of Caracas.

I really enjoyed my short stay in tropical Venezuela, where I made friends with so many young indios. Along the shore of the Bocono River, at the foothills of the Andes, my young friends had made a wooden platform on the largest branch of a huge mango tree, where we would eat mangoes and dive into the fast waters of the Bocono, swimming under water through a small rocky canyon and exit into a beautiful natural pool. The Bocono is one of the tributaries of the great Orinoco. The Bocono originates from the Andes, south of El Tocuyo, and passes by the city of Bocono, which gives it its name, and empties into the Portuguesa, which empties into the Apure and becomes the Orinoco. I never had so much fun as a child as I did in Venezuela with my indios friends, whose families were so friendly to me. There I tasted my first filets of anaconda, cooked by the mother of my friend, Joselito. Together with about twenty young indios, I helped carry a ten meter anaconda which they had killed while it was digesting a young calf it had swallowed. It was heavy and it took both of my arms around it to carry as I slowly followed my long line of friends into their village, where four men skinned it. The skin was hung to dry and it was going to be used for making purses, wallets and some of it was to be sold for money. The women filleted the meat and it was equally shared among the families of the small indios village. Joselito's mother washed the meat and placed on a large piece of banana leaves. She then mashed a few bananas and placed around the meat and sprinkled everything with ground spices. She folded the banana leaf like a package and tied it with long strings from the banana leaf and placed on top of red coals of a fire pit, where it cooked an hour. She turned the package upside down and it was clear that the green leaf had turned dark brown from the hot coal. We waited about an extra half hour and she removed the package and placed on a large flat stone. She opened it and the anaconda meat was ready to be eaten surrounded by a cream made from the mashed bananas. She spooned some meat with the banana cream on top of areppas, soft thin taco-like breads, and I had my first anaconda areppa. It was delicious and I asked Joselito if I could have another. His mother smiled and gave me a second areppa. My little friends lived in huts, slept in hammocks and swam naked. They also played games I never had experienced before. They would capture huge iguanas, five to six feet long, and tie their back leg with a small rope and race them in the main square. Men would bet money on the possible winner.

The boy with the winning iguana would receive a silver Bolivar. I just watched because I was too scared to capture an iguana of that size. They looked so fierce; but, later I found out through Joselito that they were very gentle creatures.

Before going to Venezuela I had played soccer for five years in Italy and I was considered a good player but definitely not the best. Well, in Venezuela I suddenly was the best in the state of Trujillo. A soccer scout approached my father and asked him if I could play for their team, whose youngest player was sixteen. My father was concerned for me because I was only ten years old and much smaller. He asked me if I was interested playing for an older team and I accepted the offer to play. That summer, in 1959, I scored the winning goal during the final playoff game and I became an instant celebrity. They wrote a story on the Bocono paper about me and the Bocono team had a celebration party at the best restaurant in town.

My most memorable experiences were with my father, driving to Merida up in the Andes on the Cordillera de Merida; and, hunting at La Gran Sabana, near Angel Falls. My father had promised that we would go to Machu Picchu but that did not happen: I did visit the gorgeous ruins of Machu Picchu several times as an adult. Suddenly, we had to leave Venezuela during a summer stay in Caracas when a group of rebels stormed the capital. There was fighting everywhere in Caracas and we were stuck inside a hotel, rationing food and water. It was too dangerous to go out. At the beginning of September my father ventured outside and somehow he met with friends he knew in the American Embassy. On September 20, 1960, we left in a hurry with the American Embassy on a Pan American flight bound for New York City.

We stayed in New York City for a few days, visiting my uncles and aunts on my father's side of the family. We then took a flight to Montreal, Québec, Canada, where we stayed with my uncle's family for a few months until we rented a house. Immediately, I was placed back in fourth grade because I did not speak English or French. Winter came early that year and the cold was brutal, compared to the wonderful all-year-round warm tropical weather of Venezuela. However, I enjoyed the incredible amount of snow, making igloos and tobogganing down a beautiful hill on Parc des Hirondelles, a block away from our new home. By Christmas, the only subjects that I did not receive 100%, in my report card, were in language arts; but I did have a passing grade. I quickly learned French because it is a Latin-base language and my knowledge of Italian, Spanish and Latin was and it still is excellent. By the end of the 1960-61 school year I had a perfect score of 100% in all subject, including English (speaking, reading and writing). My spelling in English was impeccable because I memorized all the words I encountered; and, after winning the spelling contest at my school, I was chosen, along with two other students to represent our school in the final regional spelling contest, which I won.

The principal of my school, Mr. Burns, who was probably the best principal I had as a student, called my parents for a conference concerning my academic success. He told my parents and me how proud he was of my diligence and success that he would promote me to sixth grade for the following school year. I was so happy because I would finally be in the proper grade. Of course, it would depend how well I would perform the following school year. My parents and I agreed and thanked Mr. Burns for his kindness and consideration. I promised them that I would do my best to keep a perfect score throughout the next school year. My parents were so proud and happy that they had a party for me, inviting my friends and a few relatives that lived in Montreal.

By then, my cosmic dreams became more daring and dangerous as I flew to other star systems and galaxies, witnessing the end of a star system being sucked into the newly formed black hole as I barely escaped the powerful forces trying to pull me in into that majestic black funnel, spinning in such incredible speed. I would wake up sweating from the terrible fight I had to endure from being sucked into my death. I was terrified but also ecstatic that I was alive. Once again, it had been one of my repeated exciting cosmic dreams.

The following school year I became more studious than ever because I did not want to disappoint Mr. Burns nor my parents, but most of all myself. I had made a promise to succeed. After school, having done my homework and having played with my friends, I would read and write short stories and poems for hours in my bedroom, sometimes past midnight. In my first report card I had a perfect score for all subjects and I proudly wore the gold medal (an incentive program at my school), which made Mr. Burns, my parents and I very proud. However, that school year there was another brilliant student, Tony Lucia, who was my top competitor in wearing that gold medal. The gold medal would be worn by the top student in the entire school population. I do recall that Tony and I evenly shared that gold medal during those four scholastic quarters.

I was so disappointed when I received my second report card at the end of January because I would have to give the gold medal back to the principal:

I did not have a perfect score in that report card; I had a 96% in English grammar. For me it was failure! I could not face my parents; I was so ashamed of myself. I could not go home. I walked to Parc Sauvé and sat on one of the cold benches beside the skating rink. By five o'clock it was already dark and I could not get myself to walk home, as I sat there, feeling disappointed and sorry for myself. As I watched young people skating on the smaller rink and a group of boys playing hockey on the hockey rink a police car, patrolling the park as usual, stopped and parked the car not far from the bench I was sitting on. Two young officers got out and walked towards me. I became worried as they stopped in front of me. One of them asked me if my name was Vincent. I nodded. He told me that my parents were worried about me and they had called the police station. The second officer asked why I had not gone home after school. I told them that I was very ashamed of my report card and of myself. The first officer asked me for my report card, which I took from my school bag and gave it to him. The officer told me that it was the best report card he had seen and that he did not see anything wrong with it. I told him that I was ashamed of the 96%; and, that it should have been 100%, just like the other subjects. The officer smiled and gave me my report card back. He told me that they would give me a lift back home. I sat in the back bench of the police car as the young officers took me back home. They rang the bell of my house and explained to my parents how I was feeling. They also told me to always go back home after school, which I promised. My parents were happy that I was safe and happy with the report card. My father told me that they were very proud of me and that next report card I would probably get a perfect score and wear the gold medal again, which I did.

During the last week of June, 1962, I was being tested by a psychologist, due to my young age, to see if I could enter Sacred Heart, a private Catholic high school in Chicago. The first test took the entire day with a break for lunch, from eight in the morning to five o'clock in the afternoon. Each part of the test took one hour and contained different subject questions, giving me one minute per question. I quickly did all the easy questions with plenty of time left to do the harder ones and a quick review of everything. I used that method for the entire test, which was subdivided by subject. The second day it was a similar test with similar questions but none were the same. I asked the psychologist if he could tell me my score; but, he told me that he would tell me how I did on Friday, the last day that we would meet. On the third day the test were quite different. That morning I had to write three compositions based on three different prompts, one of them was quite scientific. After lunch, I took two very different tests: the first was mathematics, with geometric and algebraic questions, which I really loved. The fourth day, Thursday, I was given two similar but quite different tests: the first was The Stanford-Binet Intelligence Scale and the second was The WISC. The psychologist gave me a twenty minute break; and he gave me a chocolate bar and a small bottle of milk. I ate the chocolate and drank the milk and I told him I was ready for my second test. When I saw the title, WISC, I was amused by the funny name. He told me that it was the acronym for Wechsler Intelligence Scale for Children. The test was over in an hour. I told him that both of these tests were more like games. He told me that they were both intelligence tests and quite different from the other three very long tests that tested knowledge. I asked him how I did and if I would be admitted to Sacred Heart. He told me that he would give the results to my parents. He also told me to come back with my parents the next morning, Friday, at 9 o'clock. I felt insecure and I was worried that I might not have done that well on my tests.

On that last Friday morning of June, 1962, after breakfast, my parents and I walked to the building where I had been testing all that week long, which was only a block away from our house and faced our school. When we entered the main office, the psychologist who had tested me was talking to Mr. Burns, the principal of my school. By then, I had become very worried that I had possibly done very badly. Why would the psychologist call Mr. Burns? Both Mr. Burns and the psychologist turned toward us and both came to welcome my parents, shaking their hands. Mr. Burns greeted me, telling me how nice it was to see me again. School had been over by mid-June. The psychologist approached me and told me to sit down for a few minutes while he needed to speak to my parents, who followed the psychologist and Mr. Burns to the testing room, closing the door. I became nervous and I began to wonder why I could not be included in whatever the psychologist had to say about me. I got up and went to see a poster, which had been hung a couple of feet away from the door of the testing room. I could hear the psychologist say that I was too young and something about the tests that he had never experienced before. I heard my father say that I was not ready emotionally: ready about what? My own father telling them I was not ready. Why would my father say that about me? I also heard the psychologist tell my parents that it would be better that I did not know.

Finally, I heard them walking towards me as I quickly went back to the chair and I sat down as the psychologist opened the door for my parents and Mr. Burns. My parents, Mr. Burns and I all sat on four nice chairs that had been set around a huge wooden desk, mostly covered with books and piles of papers. The psychologist sat behind the huge desk and began congratulating me on my performance on all the tests. I had done so well that he had decided to promote me to ninth grade, which I would begin in September in Chicago. Mr. Burns and my parents congratulated me and told me how proud they were of me. I finally asked how many mistakes I had made. He told me that I had not made any mistakes. I asked for my score on the I.Q. tests. He told me that it would be better for me if I did not know. He told me that my father knew and that he would tell me at the right time. The psychologist and Mr. Burns shook my hand and wished me a successful year and future. I walked happily home with my parents because I was skipping from sixth to ninth grade; but, I was still puzzled about my I.Q. When we got home, my mother began to prepare lunch, which would take at least two hours, and I began asking my dad about my I.Q. His response was that I was extremely smart and that I should be ecstatic because I was going to ninth grade in September. I smiled as I went out to play with some friends, still puzzled about my I.Q.

I spent two and a half great years at Sacred Heart H.S. in Chicago. I had great teachers; most of them had their doctorates in their fields. I learned so much and played several sports. Although, I really enjoyed those two and a half years there I became home sick. I missed my family and friends. I begged my parents to stay home after the Christmas holidays. I transferred to John F. Kennedy H.S., in Montreal, where I finished my last year. I also won three gold medals for J.F.K.H.S. in shot put, discus and javelin.

I went to Saint Joseph Teachers' College and Loyola for the next four years. It was in 1966 that I took my first Cosmology course, which I really enjoyed. Later, Saint Joseph Teachers' College became part of McGill University. I began teaching in 1970 at a new high school, after receiving a Bachelor in Secondary Education and a Teaching A Diploma from Saint Joseph.

I met my wife after Christmas of 1970; we were married on July, 1971. As I continued teaching I began a night Master's Program at McGill University. It was at McGill that I became fascinated with Differentiated Neurological

Growth between boys and girls. I did have some trouble with a couple of my professors who believed that there was no difference in brain growth between boys and girls. One of the professors insisted that I rewrite one of my papers on Differentiated Brain Growth if I wanted a good grade in his class. After, receiving my Master's degree I continued my researches on brain growth till this very day. On January 25, 1980, my wife gave birth to our baby girl, Alexandra. Two years later, on February 11, 1982, she gave birth to our son, David. I spent so much time with my children, reading to them from the very first day; and having so much fun with them doing all kinds of creative activities. I really enjoyed being a father. As a family, we had so much fun, vacationing by beautiful beaches and visiting dozens of ancient archaeological sites in Central and South Americas.

My family and I moved to Virginia, U.S.A. in 1990, where I began teaching for Prince William County Schools. I took many courses at University of Virginia in Northern Virginia and at George Mason University. I taught the sciences, mathematics, world history and geography, languages and fine arts until 2014, retiring after 44 years in education. It has taken a lifetime for me to realize that my most important goal for me is to write this book for everyone to read. Now, that I have the time I will use it to finish writing Cosmic Visions within the Microcosm of My Right Hemisphere in a clear, concise and meaningful style; and have it published by a publisher that shares my visions of the cosmos and the development of the cosmic human brain.

Finally, before I begin the first chapter, I must tell you, the reader, about the heroes in my life, whose teachings and writings have guided me into becoming myself: the person I have searched for my entire life. Firstly, Jesus of Nazareth was and will always be my number one hero for his teachings taught me about humanity, the very essence of being a good, loving, giving person. He is the one who taught us how to break away from the barbaric way of life that has ruled this planet for millennia. He willingly gave up his life, the ultimate sacrifice, to show us that love and peace can, one day, save this beautiful planet. As a young artist, Leonardo Da Vinci became my second hero. His art inspired me to draw, paint, and sculpt every day of my life. As a scientist and inventor, Leonardo clearly was five hundred years ahead of his time. His visions were very futuristic and prophetic. As a young scientist, Albert Einstein, became my third hero for he opened up, to our modern world, the window to our universe. His genius and visions allowed science to develop exponentially in all directions. Finally, my only female hero was and still is Mary of Nazareth, the mother of Jesus. For me, Mary was and will always be the very essence of motherhood. Her love for her son and the pain and suffering she had to endure are the evidence of all the love and pain that all mothers experience in their lives. Her pain and suffering also symbolize the brutality and suffering that women, throughout the history of mankind, have experienced to this very day. There are millions of women in today's modern world who are victimized, enslaved, traded and sold as slaves, beaten and killed by men who pretend to be righteous and religious, but are no better than pimps and murderers.

My spring, summer and autumn of my life have long been gone, leaving me with so many beautiful memories of people and places that have had profound meaning in my life. Now that I am in the winter of my life I feel the urgency to have this theoretical treatise on the cosmos, black holes and the microcosmic universe of the human brain published before I leave this world. Lately, as if I have returned to my early years of my life, the cosmic visions in my mind have become more intricate, clearer and more meaningful. Every night and day these visions haunt my very existence. I have decided to make this a concise and readable treatise so that most people, including some very young brilliant minds, will be able to see and understand my cosmic and microcosmic visions.

Vincent L. Di Paolo (2018)

"Well we all shine on
Like the moon and the stars and the sun…
Why in the world are we here
Surely not to live in pain and fear
Why on earth are you there
When you're everywhere…" John Lennon (1975)

CHAPTER 1

CREATION: THE FIRST BIG BANG

As a scientist I cannot divorce myself from the concept of a Supreme Being; however, trying to prove the existence of a Supreme Power is as elusive as trying to prove any of the theories on the universe. I know that most cosmologists and physicists do not believe in a Supreme Being; and, some believe that our universe and the entire cosmos developed from nothing: I fail to understand that false logic! Whether one believes in a Supreme Being or not it does not change the presence of a universal power everywhere throughout the cosmos. If the reader wishes to call that universal power God, Dios, Dieu, Allah, Yahweh, or any other name it is fine with me on the condition that you, the reader, does not imagine the Supreme Being as a person, male or female. The closest symbol that I can think of which might rightly symbolize the Supreme Being is the Yin and Yang, the negative and positive forces that control the entire cosmos. Thus, in this book I will call this cosmic power *the Supreme Power, the Supreme Matter* or *the Supreme Being.*

Although most theoretical physicists tend to agree that the first big bang occurred approximately 13.7 billion years ago, I will stretch that very beginning to a period between 15 to 20 billion years ago or probably more, which I will illustrate the reasoning for that extra time. For now, by using our imagination, let us go back to that very beginning of time. At the very beginning, the cosmos was so much smaller than what it is now *(where possibly each cubic millimeter held more than 100 trillion hydrogen atoms)* and its only inhabitant was the Supreme Matter, which held all the DNA that will ever exist and develop and the power to create all the elements of today's and future universes, coupled with all the intelligence that will ever exist in the entire expanding cosmos. The Supreme Matter decided to share all of Its universal DNA and all of Its cosmic intelligence by exploding Itself, creating the Initial Big Bang, and thrusting all of Its DNA and intelligence into every possible direction of an extremely malleable cosmos: It was all hydrogen atoms with the power to develop into star systems, all elements, galaxies, and universes with life and intelligence everywhere. To imagine an image of that First Big Bang you must visualize a willow or a serpentine firework, where it explodes in the dark night sky and then breaks into smaller pieces, creating their own smaller explosions. The force of the First Big Bang was so powerful that it began to inflate the cosmos very much like a super gigantic balloon; and as the smaller broken parts of the Supreme Matter flew across the globe-like cosmos and exploded into their individual big bangs they stretched the cosmos in all its global directions, where each smaller big bang had been programmed to develop into its own universe. In the first few minutes after the very first Big Bang the temperature of that baby cosmos must have reached at least a probable one billion degrees Celsius or more fusing some of the original hydrogen into the first original helium. The instantaneous super rapid expansion of the baby cosmos slowed down, cooling down the temperature as it reached the sufficient production of helium in each new healthy star, leaving the remaining hydrogen as *Lord* of all the elements. The constant super high temperature had been programmed to last only a very short time within the young expanding cosmos, creating enough helium needed for the future creation of the very first young and healthy stars; and, within each healthy young star was the programmed capability to create all other elements needed to begin life within their respective fertile planets.

One of the smaller big bang was the beginning of our very own universe. After hundreds of million years, its intelligence inhabited an infinitesimal number of subatomic particles. Each and every subatomic particle had been given a super protective coating that I am naming *alexion* ©; and, as they joined together they became the first atomic nuclei in nebulae and in super mega protogalactic clouds, eventually developing into *protostars* and *protogalaxies*. *Alexion* was given the synergic power to protect each and every subatomic particle, that will ever be created in the cosmos, and to withstand any temperature anywhere in the cosmos and to protect the information that each subatomic particle held. With time the protogalactic clouds began to spin around their centers which eventually developed into galactic disks, each spinning on an ever increasing axis with increasingly powerful gravitational centers everywhere in our universe. The very first young stars eventually were able to fuse hydrogen into helium and into carbon, oxygen, beryllium and all the other elements that we know today and will discover in the future. Our Big Bang and all the other big bangs grew exponentially and they are still growing at an exponential rate. All of the other universes shared the same information and DNA; but, each one developed differently and uniquely. All universes are very similar and related but each one is unique in how it developed with a different possible combination of information. There is a high probability that all universes share ninety-nine percent of all cosmic information; but, it is that one percent difference that makes each universe unique. Although a general set of laws of physics are most probably shared by all universes each universe most probably has its own personal subset of laws of physics that can be applied only to that individual universe.

During the first three to five billion years, infinitesimal numbers of hydrogen atoms got together to form the first generation of healthy and productive stars, which greatly illuminated the young spherical cosmos. Our universe and the entire cosmos cannot be flat! Our universe is a globe, most probably an elliptical globe, pressured by the gravitational forces of neighboring universes *(babies and large mature ones)*, distorting our spherical universe as other universes are distorted by our universe's gravitational force. The entire cosmos is constantly growing spherically.

During the second five billion years, some massive stars broke into medium and smaller stars, spinning around their newly created axes. Stars shot out parts of the fiery masses as they began orbiting their respective stars. Most of those fiery sub-masses shot out smaller parts, gathering early cosmic dust as they began to spin on their newly acquired axes, creating planets and moons as they began to cool down. It is probable that during this period of the cosmos *(one to five billion years after the very First Big Bang)* some of those first planets had all the qualities to create life: proper distance from their stars, water, carbon, hydrogen, nitrogen, oxygen, sulfur, phosphorous, calcium, potassium, iron, magnesium and other elements needed. It was probably during the second period *(five to nine billion years after the First Big Bang)* that the first fertile planets held the first populations of intelligent beings, some very similar to us today, others might have developed differently from us but were as intelligent or more intelligent than we presently are. Those first intelligent and fertile planets do not exist anymore as the first stellar and galactic black holes began to recycle star systems creating new productive nebulae. Black holes were and are programmed to be the recycling machines of the cosmos as they break down dead star systems into their atomic and subatomic particles; and, as those subatomic particles are launched out of both ends of the black holes they join existing nebulae or begin new ones, eventually beginning their long process of reassembling into new star systems. The first galaxies with intelligence were most probably formed during this second period of cosmic expansion *(five to nine billion years after the First Big Bang)*. Some of those galaxies do not exist in the same state as some of them have merged with others, blending a significant percentage of their star and planetary systems into mega galaxies.

In creation, which is ongoing and never ending, entire sections of galaxies are destroyed as new ones are created. Old star systems and complete galaxies are recycled through gigantic galactic and cosmic black holes, which eventually reassemble in the galactic nebulae as new stars systems and eventually into new baby galaxies.

Fig. 1: (NASA-HS20: telephotos taken by Hubble Telescope)

Pure energy never dies; it is recycled into new states. Everything in our universe is recycled into something new, including us. As our universe expanded so did all the other universes, stretching the elasticity of the cosmos in all directions. Each universe pushes into other universes' spheres, depending on its gravitational force and need for expansion.

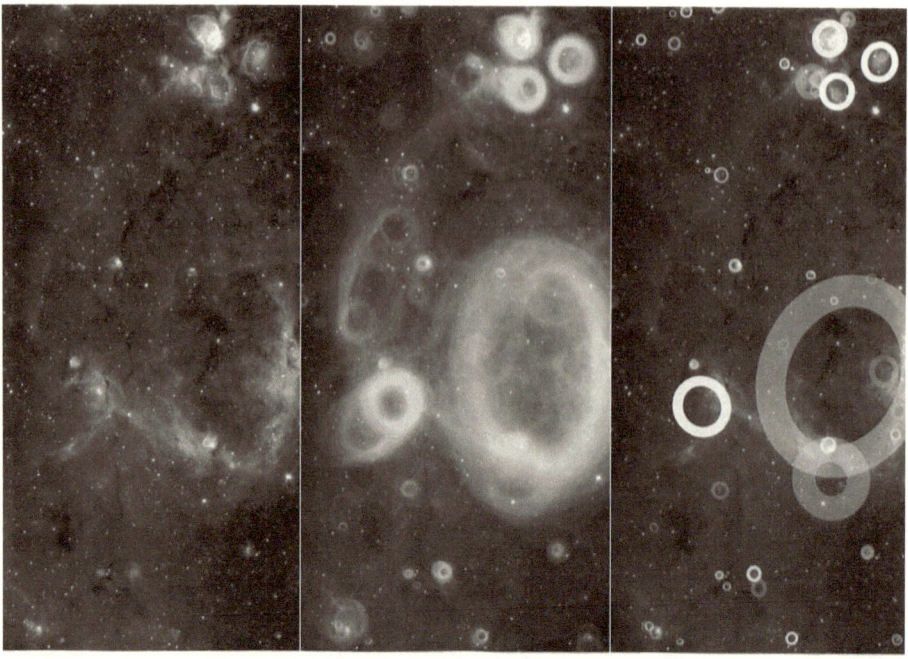

Fig.2: (NASA- Telephotos by Spitzer Space Telescope)

The malleability and flexibility of the cosmos allows the compression of a smaller universe's space: as a larger universe expands it pushes its outer limits onto the boundaries of a smaller one. When the larger and expanded universe reaches its maximum cosmic boundaries, as it nears its end, its massive cosmic black hole prevails by swallowing it entirely breaking it down to its original subatomic particles, reducing its space and allowing the expansion of smaller neighboring universes as a new baby universe begins to develop from the newly formed massive cosmic nebula to replace the pulverized old one. This give-and-take of the cosmos allows universes to grow to their maximum capacity only to be recycled to warrant the birth of baby universes and the expansions of younger and smaller ones.

Creation goes on forever, programmed by the Supreme Matter to destroy and create new universes from the *ashes* of old burned-out ones. Universes are most probably finite *(up to hundreds of billion years or more)* as the largest cosmic black holes destroy and reduce them to their subatomic states, allowing the reassembling of their atoms and eventually becoming new baby stars, galaxies and universes, pushing the fabric of their space-time as they begin to expand anew. Within our progressively and ever changing universe creation continues and evolution is a vital part of its progression. Without black holes creation and evolution would come to a quicker end; black holes recycle cosmic bodies perpetuating creation and evolution of all living things. For us, as the inhabitants of this beautiful living planet, the most important object of creation is the development of the human brain, which I call the *young cosmic brain*.

The Milky Way, according to most theoretical physicists, is between 13.4 to 13.7 billion years old. Our sun, a medium star of the Milky Way, is approximately five billion years old. Since our planet Earth is a product of our sun it is about 4.5 billion years old; and, our moon most probably developed from a fiery shoot of our young fiery globe. If our galaxy is between 13 to 14 billion years old, then our universe must be at least between 15 and 20 billion years old or more for it took a few billion years for the first galaxies to begin to form. There are between 250-350 billion stars in our galaxy with hundreds of billions of planets. The probability that our planet is the only living planet with intelligence is next to zero. The probability that for every million planets in our galaxy there is one living planet with intelligent life is very good. That would give us hundreds of thousand planets with intelligence. If you, the reader, feels that these are too many planets with intelligent life in our galaxy; then, the probability that for every billion planets there is only one with intelligent life it would give us hundreds of Earth-like planets in the Milky Way. But you, the reader, just like some of my former students, might say that planet earth is the only planet in our galaxy with life and intelligence. Let us take this minimalist probability that only our planet has life and intelligence *(which is mathematically highly improbable)*, then would the reader agree that there is only one living planet per galaxy that has life and intelligence. Presently, we know that there are more than three hundred billion galaxies in our visible known universe *(we can only see one third of our entire universe)* and, at one intelligent planet per galaxy, that would still give us about three hundred billion living planets with intelligence. However, if we could see our entire universe we might be able to see a trillion galaxies or more; and, that would give us at least a trillion planets with intelligence *(based on one planet with intelligence per galaxy)*. By using mathematical probability, I state there is the possibility of hundreds of thousands of living planets with intelligence within our Milky Way. Remember, at creation, the Supreme Matter released all the DNA and intelligence that the cosmos will ever hold in all directions. It is impossible that we are the only planet with intelligence! There are older living planets out there in our galaxy and in all other galaxies and in all other universes that have intelligence. We know that we, the people of this earth, are still at an earlier stage of neurological development; and, we know that we have so much more to explore as we continue to develop and grow neurologically. In our universe there are probably sextillions of planets *(1,000,000,000,000,000,000,000)* that are similar to earth and that can possibly sustain life. There is a high probability that quintillions of those planets have intelligent beings. We are one of those living planets.

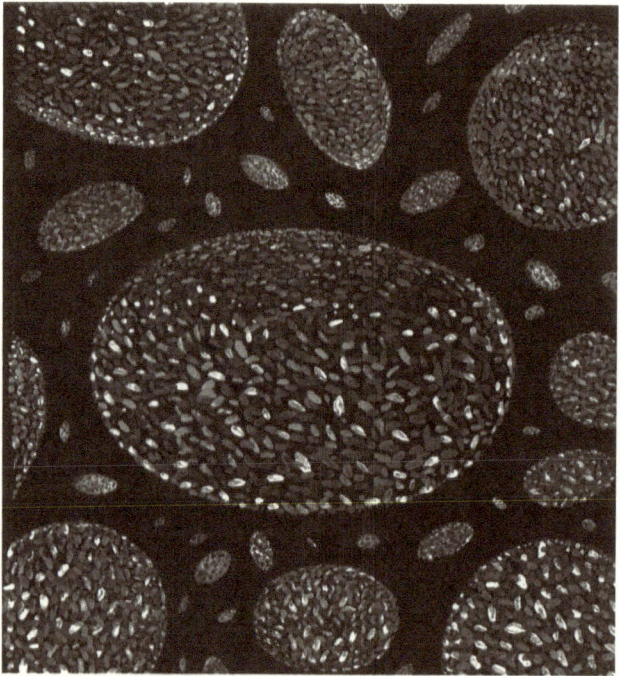

Fig.3: (Our universe surrounded by other universes) Vincent L. Di Paolo, 2016.

The probability that there are much more advanced intelligent beings within our galaxy, within our universe and within the cosmos is very high. But, there is also a very good probability that there are equal intelligent beings or intelligent beings at earlier stages of intelligence out there. That would put us somewhere in the first 12% of the intelligence bell curve. Can we possibly travel to another earth-like planet of another universe? I believe we will never be able to travel outside of our universe because of possible buffer boundaries, protecting and holding each universe, which would never permit us to penetrate through. We would most probably be disintegrated if we would ever manage to travel out of our universe and into another. Thus, we are bound within this universe with its unique laws of physics, and most probably within the Milky Way for millennia. It is presumptuous of us to believe that there are no intelligent beings in fertile living planets of other universes and in fertile living planets in the trillion galaxies within our universe and in the millions fertile planets within our own Milky Way.

It is very clear and evident that space is curved and global and that time differs, ruled by each individual gravitational force! We have proof that all cosmic bodies are global and galaxies are global or elliptical dishes holding a *smorgasbord* of billions of star systems, with planets and moons. The space between all celestial bodies is there to protect and guarantee growth and privacy. If we want to call this space *black matter*, then black matter is the guardian of all cosmic bodies, guaranteeing protection and long life of most cosmic bodies. Dark matter is the *cartilage* in between each and every cosmic matter, including the space between atoms in the micro cosmos.

If our Universe was a direct result of the very first Big Bang then our own big bang and all the other big bangs immediately and simultaneously followed that initial one: the original information and DNA were equally or nearly equally shared. We can conclude that fertile planets, life, and intelligent beings are everywhere in each galaxy of our universe and in each and every galaxy of all the other universes within our ever expanding cosmos. Our universe most probably belongs to a cluster of universes that revolve around the center of the cosmos.

Within our galaxy, the Milky Way, we share with other intelligent planets a high percentage of the same subatomic particles *(from billions of recycled star systems)*, which made up the nebulae around it five billion years ago. Thus, we share a meaningful percentage of the same type of subatomic particles from those billions of old star systems that were recycled, sharing similar foundations and galactic information within our neurological systems. It is very probable that one of those far away galaxies might be an exact copy of our galaxy, with the same number of star systems, with a star and planetary systems called the Solar System containing an exact copy of planet Earth, and another Vincent L. Di Paolo is writing the same book that I am presently writing.

Finally, I truly believe that *Dark Matter* is extremely vital to the lives of all cosmic bodies. Dark matter is at the very least 95% of each universe, cushioning each and every cosmic body *(from the smallest asteroid to the largest star, and from the smallest galaxy to the largest cluster of galaxies)*. Thus, dark matter is the *cartilage* of the cosmos, protecting all visible matter and even subatomic particles in each nebula of each and every galaxy. There is more dark matter between star systems than between planetary system. The larger the planets the more dark matter is needed in between them to guarantee a cushioning from the powerful gravitational forces of each one. The vast amount of dark matter between star systems guarantees the life of those stars. The amount of dark matter between galaxies is astronomical and the amount between clusters of galaxies is cosmic, protecting all cosmic bodies and permitting their longevity. However, if dark matter is reduced between two cosmic bodies they eventually incorporate or obliterate each other depending on their velocity of their encounter.

CHAPTER 2

GRAVITY: HOW IT RULES THE COSMOS AND OUR VERY LIVES

Gravity is the cosmic force that controls all the various bodies in the cosmos, from the smallest asteroid to the largest galaxy, creating cosmic geometry and mathematics as stronger gravitational forces pull smaller ones, expanding and warping the very fabric of space-time. The gravity of a mega star, with a gigantic iron core, is so powerful that it slows down time and the space it occupies, greatly warping its space-time; however, its quick fusing of its hydrogen into helium through the use of carbon, nitrogen and oxygen shortens its life.

The most powerful gravitational forces that govern galactic and cosmic black holes slow down time to nearly a standstill, especially during their cosmic jobs of breaking down matter to its smallest subatomic particles. The Supreme Power reigns in all the cosmic gravity; thus, all of the cosmic gravity *(from the very smallest to the largest)* have become the offices and the working centers of the Supreme Power, constantly creating cosmic mathematics to rule each and every cosmic body, each subatomic particle, each quark, each neutrino and each and every photon. Gravity not only controls the entire curved cosmos but it also continuously keeps creation ongoing through the central forces generated by its various black holes and later by reassembling each and every subatomic particle *(that are constantly recycled from old star systems)* into new nebulae, which eventually nurse those mixed subatomic particles into new baby star systems and baby galaxies.

Gravity throughout the cosmos is negatively charged while mass is positively charged. The ever expanding cosmos continues its creation, constantly borrowing from its gravitational forces. Creation is ever present in each and every galaxy as old star systems are recycled inside their galactic black holes and new ones are formed in all galactic nebular suburbs, contributing to the constant expansion of the cosmos as the baby galaxies grow into adult galaxies, like our Milky Way, and even into super mega galaxies, like the Andromeda Galaxy. Although space-time is everywhere throughout the cosmos, Earth's space-time is unique to our planet and its rules cannot be applied anywhere else in our galaxy or in our universe. Earth's space-time works only for us; outside our planetary system *(which includes our Moon, man-made satellites and Space Station)* our space-time is invalid. Space-time is relative to where we are *(Earth)* for we have created the time we observe based on twenty-four time zones *(to facilitate our daily lives)*, and Earth's rotation on its axes and its speed of its revolution around the sun. Each and every other planet of our Solar System has a different speed of rotation and revolution, each creating a unique space-time. Each and every cosmic body has its own variation and unique space-time. Also, space-time is bound to the rules set by each cosmic gravitational force and speed, warped and differentiated by each and every gravitational center: it is relative to each specific cosmic body in an endless space where time *(as we know it)* cannot be the rule for it would impose a quicker finite existence to the forever expanding cosmos.

Dark energy rules the entire cosmos, within each individual gravitational center, from the smallest to the largest degree of power: from the earth's gravity *(which keeps us safe)* to the super gravity that keeps gigantic clusters of thousands of galaxies spinning at unbelievable speeds around the very centers of their respective universes. It is highly probable that there are nearly 20 sextillions (20,000,000,000,000,000,000,000) healthy functioning stars in the part of the universe that is visible to us; and, there are possibly at least sixty sextillions of normal and healthy functioning stars in our entire universe, together with hundreds of sextillions of dwarf stars, older giant stars, neutron stars, failed stars and so many other types of mega stars or lesser stars, all traveling at cosmic speeds controlled by gravity. As gravity accelerates cosmic objects it also accelerates the time of our lives, eventually bringing us to our own very end of life. Gravity keeps us alive on this beautiful light blue planet; but, its great force slowly consumes our bodies' organs and all of our cartilage and eventually kills us.

    The rotation of a cosmic body, on its axis, causes the gravitational force of that body. The gravitational force created by the earth's rotation, spinning at 1,674 kilometers per hour, is strong enough to hold its population of people, animals and all the objects we have created in place and strong enough to hold the moon and all the man-made satellites revolving around it. The rotation of the sun on its axis, at more than 7,000 kilometers per hour creates a gravitational pull so great that all planets as far as 100 AU *(149,600,000,000 kilometers= 1AU)* and possibly farther revolve around it while the entire Solar System travels around the Milky Way at approximately 864,000 km/hr. The gravitational force of the Milky Way pulls all of its stars closer to its center with every revolution as the entire galaxy is revolving around the axis of the cluster of galaxies it belongs to at approximately 2,124,000 km/hr. Our cluster of galaxies *(made up of possibly over a 100 galaxies in a super group of more than a 1000 galaxies)* **is revolving at much greater speed around the very centre of our universe, where the largest super massive cosmic black hole** *(10-20 billion times the mass of the sun)* **controls all of approximately a trillion galaxies, or possibly more.** The closest galaxies *(probably the oldest)* to the super gargantuan cosmic black hole are sucked into it to be recycled into an unimaginable number of subatomic particles *(each carrying one bit of information of those galaxies)* and then spewed into the cosmos to eventually compose the largest nebulae, which with time will reassemble into billions of new stars to begin new galaxies with mixed information, ready to create life in earth-like planets and eventually intelligent beings with brains very similar to ours, for that is the way we developed and continue to evolve.

    As our Solar System began its first revolution around the Milky Way, after it was created by the re-assembling of subatomic particles from recycled older stars in one of the suburban nebula, it took over 500 million years. By the second revolution, our Solar System took less time as the powerful galactic gravitational pull drew it closer towards the center of our galaxy. With each revolution, our Solar System took less time as it got closer and closer to the gargantuan galactic black hole that rules our galaxy. According to my calculations, we are probably on our 17th or 18th revolution around the Milky Way as we have reached about half way within the radius of our elliptical revolution. Presently, it should take us only about 205 million years, at approximately 864,000 kilometers per hour, to go around the galaxy. In the future our Solar System's projected revolution will get smaller and smaller, reaching faster and faster speeds as the super powerful gravitational force of the Milky Way's mega black hole draws us closer and closer until it reaches its central bulge. By then the sun will have become a red giant star and it will have engulfed all of its planetary system. Ten billion years will have passed and there will be no memory left of planet earth and its history. As our Solar System, along with millions of other old star systems, will be pulled into the Milky Way's gargantuan black hole to be pulverized into an astronomical number of subatomic particles, each one holding one bit of information about our Solar System, our planet and each one of us, as they are spewed into the cosmos and eventually drawn back to the suburbs of nebulae around our galaxy.

In my vision, I see the oldest galaxies being drawn closer and closer to the center of our universe where the biggest and most powerful **cosmic black hole** *(10-20billion times the mass of our sun and possibly thousands of times bigger than the largest galactic black hole)* **controls our entire universe with the most magnificent dark energy's gravitational pull that reaches the new and most distant galaxies, which dwell in the farthest regions of our universal sphere. The universal black hole is spinning at the fastest speed next to the speed of light. It is able to pulverize complete galaxies that have existed for twenty billion years or more by creating winds that move over 1,000,000 kilometers per hour and create a temperature of more than a billion degrees Celsius, and spew out an unimaginable number of subatomic particles** *(each* *holding one bit of information, protected by alexion from that imaginable heat)* **out into the super cold cosmos to form the most beautiful and colorful nebulae, which with time** *(millions of years)* **will begin to reassemble into millions and millions of new star systems and eventually into new galaxies and millions of earth-like planets with intelligent life.**

"We are stardust, we are golden,
We are billion year old carbon,
And we got to get ourselves back to the garden." Joni Mitchell (1970)

CHAPTER 3

SCHEMATA OF A GALAXY AND THE FUNCTION OF ITS GALACTIC BLACK HOLE

The Milky Way and all other galaxies are super mega-factories of stars as the dark energy within their galactic black holes *(cosmic super recycling machines)* plays a major role in the lives of old stars by constantly pulling them in and recycling them into astronomical numbers of subatomic particles; and, then spewing them out to join other subatomic particles in the nebulae of their respective galaxies. There, once cooled down by the extremely cold cosmos, the subatomic particles join other subatomic particles of other recycled stars as they begin to reassemble their information into new stars. Thus creation is ongoing in each and every suburban nebula of each and every galaxy.

In this chapter I will illustrate what I envision are the functions of galaxies both through language and art. My illustration of a galaxy will reflect the activities within our beautiful Milky Way and all the other galaxies that exist in our universe and in other universes. As I have mentioned above, all galaxies can be easily divided into three equally important parts: the main visible part is the elliptical disk-like track on which hundreds of billions stars travel around at super speeds; at the center of each galaxy there is a galactic black hole *(a cosmic super machine)* which recycles all the older stars surrounding it *(which have become super giants)* that have reached their last cycle, forming a central bulge, before they are sucked in from both sides of the galactic black hole's portals; and outside the track are immense nebulae *(galactic clouds)* that are constantly formed or reshaped by the cosmic number of subatomic particles that are spewed out of the black hole's upper and lower mouths into the nearby cosmos. The powerful gravitational force produced by the super speed of the black hole spinning on its galactic axis at more than 1,500,000 km/hr. pulls back all of the spewed subatomic particles onto the suburbs of each galaxy to form super gigantic nebulae which become the nurseries of newly born stars from the cosmic mixture of those same subatomic particles *(which are the product of billions of pulverized old stars)*.

Fig. 4: (Painting of a galaxy, similar to the Milky Way) Vincent L. Di Paolo, 2016

Galaxies are spiral, elliptical and several other types of cosmic *dishes* containing hundreds of billion stars racing around their galactic tracks. Their galactic black holes are super recycling *machines*; and, their suburban nebulae are *factories* and *nurseries* of new baby stars, which are created by the astronomical numbers of subatomic particles from old recycled stars as they finally begin their long journeys at the very edge of each galaxy. Each new star system gets closer to the galactic center with each new cycle. After about five billion years those new stars become middle aged *(just like our sun)* as they reach near the midpoints of their elliptical radii *(cosmic pi varies from lower to higher than 3.14 as their track gets squished by various galactic gravitational push and pulls)*. It will take another five billion years for those same stars to reach the very edge of the central bulge, covering the powerful galactic black hole *(by then many have become super red giants and have totally engulfed their planetary systems)*, to be pulled into its super grinding winds of the galactic recycling *machine* which will pulverize those old stars into astronomical numbers of subatomic particles *(each and every subatomic particle holds one bit of information from the star system it belonged to)* and finally they are spewed into the nearby cosmos, only to be regrouped into nebulae around their galactic suburbs to eventually become new stars. Nothing is wasted; every single particle and information is reused as creation goes on each and every nanosecond. We, the intelligent humans of planet Earth are the product of mixed subatomic particles from billions of recycled star systems *(that happened approximately five billion years ago)* with mixed information and intelligence from each and every one of them.

CHAPTER 4

BLACK HOLES: SUPER MACHINES PROGRAMMED FROM THE BEGINNING OF TIME TO RECYCLE STARS, GALAXIES, AND UNIVERSES

In an ever expanding cosmos *black holes* play a necessary and dynamic role in the constant creation and evolution of star systems, galaxies and universes. In the past two decades my visions of black holes have evolved to highly sophisticated and necessary components of our universe: black holes are the most advanced cosmic *machines* that have been programmed from the very beginning of time to recycle burned out star systems, older galaxies and terminal universes into new stars, new galaxies and baby universes. Dark energy within each black hole rules their purpose and functions. Without black holes our universe and all its galaxies would stop expanding and eventually come to a much quicker end. The very presence of dark energy *(negative force)* in each and every type of black holes guarantees a recycling, a continuity of creation and ever growing expansion of the cosmos.

According to Chandrasekhar's amazing cosmic mathematics all stars that are greater than 19.59M *(M= mass of our sun)* are programmed to develop their own stellar black hole as a guarantee that all its energy, all of its elements and DNA containing all possible intelligence and information and potentiality as a star system, with planets, moons and possible life is recycled. The death of each large star system by its programmed development into a stellar black hole guarantees all the subatomic particles needed for the birth of a new star system. Depending on the size of each star system, stellar black holes will differ in size throughout the cosmos. The bigger the star and its system the bigger will be its black hole. No matter the size of larger stars, the mechanics of all black holes follow the same rules with variations based on their individual axial rotating speed. For example, the speed of the rotation of the central axis of our sun is approximately 7,000 kilometers per hour at its core and about 33 % reduction at its poles. I believe that the centrifugal speed of the sun will speed up at the end of its life; and, that is most probably true for most stars in the cosmos. The increase of its centrifugal speed is directly related to its expansion to facilitate the recycling and pulverization of each planet, moons and other lesser bodies that belong to it. However, the sun will most probably not create its own black hole for it will not reach the necessary production of iron in its core in order to implode.

Stellar black holes exist within each and every galaxy, each one programmed to recycle individual star systems. At the end of their lives, older giant stars implode after reaching their maximum expansions, creating their own personal black holes. Those stellar black holes continue to travel along their trajectory at the same speed, sucking in any stellar objects that might come their way. When a stellar black hole has done its job in disintegrating its entire star system it will eject all of its subatomic particles out into the section of the galaxy it belonged to join the nearest nebula or form a new one: this final part of the job of a stellar black hole ends its life, totally disappearing from view, leaving a new nebula *(containing all of the subatomic particles, DNA and intelligence of that recycled star)*. It is very difficult for astronomers to detect stellar black holes for they have become a small fraction of the old star systems they developed from. In Figure 5, illustrated in five steps I will explain how larger stars more than 19.59 times our Sun implode, at the very end of their expansion, and become stellar black holes. In the first sketch in Figure 5, a large star has reached its maximum global growth, burning all of its planets and moons. The second sketch shows an implosion at its weakest point, causing a powerful suction as it begins its formation of a stellar black hole. The third sketch shows that it has become a full stellar black hole after fusing a staggering number of iron, making it extremely heavy with a super gravitational pull, disintegrating the entire stellar system into a stellar number of subatomic particles. The fourth sketch shows when the black hole has done its job of pulverizing its entire stellar content it will spew all of the subatomic particles out of its top and bottom poles, creating a corona of stardust.

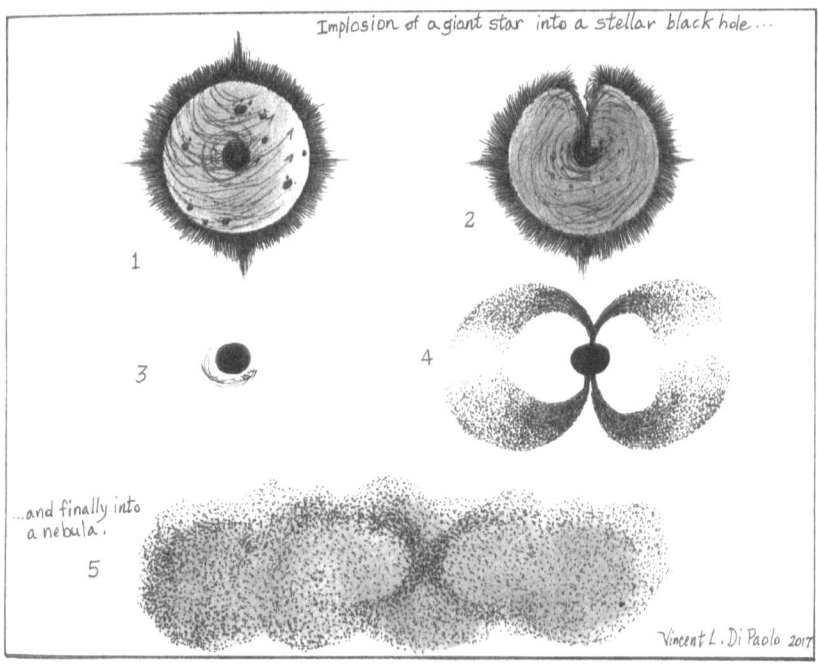

Fig. 5: *(Five sketches, illustrating the Implosion of a stellar system into a stellar black hole and finally into a nebula) sketches by Vincent L. Di Paolo, 2017)*

The fifth sketch shows that stellar number of subatomic particles has formed a nebula as the black hole disappears as its life and purpose have been accomplished.

Galactic black holes are so enormous and powerful, generating such potent and efficient gravitational forces that control hundreds of billions star systems and mega nebulae spinning around each respective galaxy. The most gargantuan cosmic black holes recycle entire universes and they are the very centers of their universes. A universal black hole can be more than twenty billion times larger than our entire solar system, generating an enormous gravitational pull to keep a trillion galaxies spinning around it at unbelievable speed as it recycles entire galaxies that finally reach its very edge. That enormous and most powerful gravitational pull generated by an unbelievable speed of the universal black hole, as it rotates on its universal axis, keeps a trillion galaxies or more forever spinning closer and closer to its center for its most important role is to recycle complete galaxies and eventually create new ones from the astronomical numbers of subatomic particles produced by each recycled galaxy. Thus, creation is ongoing, lasting hundreds of billions and possibly trillions of years.

Black holes do not exist by chance and they are not chaotic accidents of the cosmos. Stellar black holes have been programmed to develop to do their own recycling jobs by the Supreme Matter at the very beginning of time in each and every star larger than 19.59M (*smaller stars like our Sun do not develop into black holes*) in each galaxy and universe; and, they are the very last action of each large star. Stars that do not develop their own black holes will eventually be permitted to enter their respective galactic black holes at the end of their lives after their final revolution. All the information, intelligence and energy of each and every star, of each and every galaxy, and of each and every universe is not lost: it is recycled into its subatomic particles, each holding one bit of information protected by *alexion* (*a protective coating that is programmed to withstand any temperature or any speed*), **which will be used to reassemble and create new stars, new galaxies and new baby universes. Nothing is lost; everything is reused to guarantee a continuum of creation and expansion throughout the cosmos.**

There are stars who fail to mature because their inner mechanics do not function properly due to so many possible ill factors found, at times, in chaotic cosmic accidents throughout each galaxy and universe. You, the reader, might ask if stars can get sick? The answer is, simply and accurately, *yes*! How is that possible that a star or galaxy can get sick? Each and every creation, from a quark to a galaxy, from a blood platelet or neurotransmitter of a person or animal, has a time limit to its life. Yes, that is true; but, you might tell me that we are constantly *(through science)* increasing the duration of our lives. True, but even if we, as humans, might one day live a thousand years it would equal a nanosecond in comparison to the life of a universe. Stars live billions of years if they are healthy; but, if their lives are affected by some cosmic illness *(ex.: not able to create enough helium through fusion)* **their very lives are changed and they never develop into full and healthy star systems. So many stars remain small, dwarfs, and they will not be able to develop their inner mechanisms to mature into a healthy star. They will eventually be recycled by the galactic black hole of the galaxy they belong to. The following four illustrations that I have painted will visually illustrate the extremely important function of galactic black holes. These four theoretical performances of galactic black holes are constantly present in my right hemisphere, clearly showing their grandeur in their constant and most significant function of everlasting creation.**

In Figure 6, below, you see an average galaxy *(very similar to our Milky Way)* sliced in half to permit a view of the function of its galactic black hole.

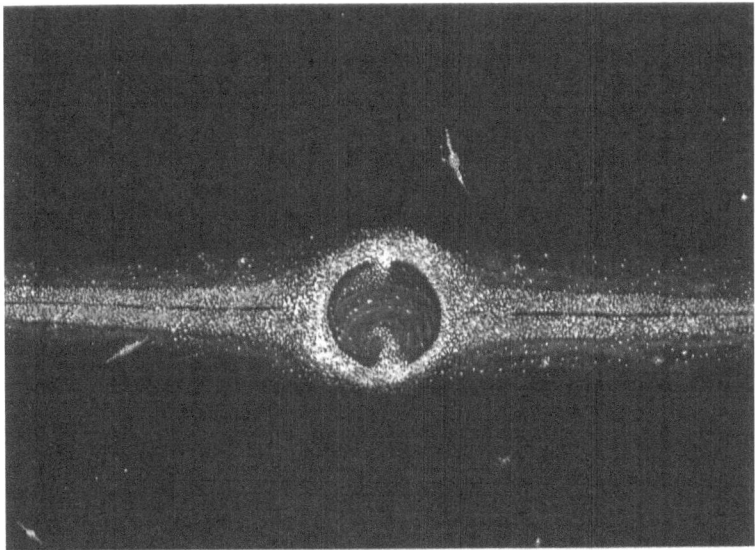

Fig. 6: (Painting of a galaxy sliced in half, showing its galactic black hole in its first operation of its total function: pulling in millions of old stars, waiting to be recycled) By Vincent L. Di Paolo, 2017.

Only by slicing a galaxy one can envision the enormous black hole present in the center of the galactic nuclear bulge, which is made up of billions of older star systems that have reached the galactic nucleus and have become part of it. After an average of ten billion years traveling at unbelievable speeds *(At their beginning young stars travel hundreds of thousands km/hr.; while the speed of our solar system is approximately 864,000 km/hr., at the center of its galactic radius. Stars slow down in their last few revolutions before they are permitted to enter one of the two portals of their galactic black hole.)* older star systems reach the central bulge that enshrouds the upper and lower portals of the gargantuan galactic black hole as they wait until they are finally permitted to enter its huge gates. Once millions of oldest star systems enter the enormous galactic super grinding machine their lives are about to come to a quick end as its winds will possibly reach 298,387 km/s., just under the speed of light, and its temperature might be a possible billion degrees Celsius. No one knows for sure how many old stars are allowed to enter during each pulverization; but, we can imagine

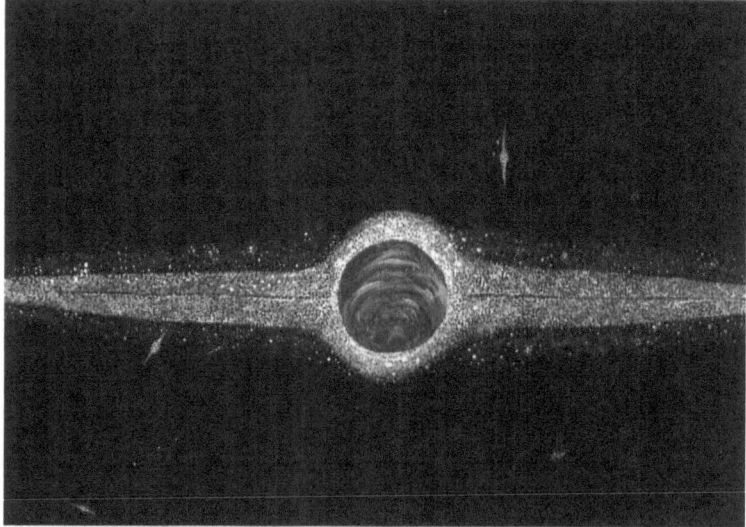

Fig.7: (Painting of a galaxy sliced in half, showing the second operation of the function of the galactic black hole with its portals closed as its super winds and temperature will eventually pulverize all of the captured old star systems.) By Vincent L. Di Paolo, 2017.

millions of those old mega stars enter before the portals close. Each opening can easily last millennia of earthly years *(a small time opening in the life of a galaxy)*, permitting millions or possibly a few hundred millions of old giant stars to be recycled. At this point nothing that enters is permitted to exit as those old stars are swept by interior winds possibly reaching close to the speed of light as the galactic black hole *(of our Milky Way)* reaches the culmination of the total pulverization of those old star systems. Also the pressure of hundreds of millions of old stars constantly place such a force on the portals that nothing would be able to come out as more old stars keep pushing onto the polarized mouths.

In Figure 7, the polarized portals have closed shut and the spinning of the winds inside the black hole can possibly reach just under the speed of light as the millions of old star systems spinning around in a deadly and final stage of their lives. The temperature inside a galactic black hole filled with millions of old stars might reach well over 1,000,000,000 C° and thousands of earthly years might pass when all of those old star systems will have been pulverized into a cosmic number of subatomic particles. Each particle retains one bit of information from the star system it came from; and, each subatomic particle is protected by *alexion (©2017)* from the unbelievable heat and speed. The super speed and heat of the interior of the black hole intensifies and it reaches its culmination as it gets ready to erupt, containing a cosmic number of subatomic particles, holding all the intelligence of each and every old recycled star systems.

In Figure 8, the galactic black hole has completed its function of total pulverization of the millions of old star systems it sucked in for that particular job. The pressure from the intense heat and speed of its interior winds continue to spin the astronomical number of protected subatomic particles. The billions of old star systems around the north and south poles of the central nuclear galactic bulge spread out to free the spaces that cover both portals. As the portals open two galactic geysers on each pole erupt with speeds well over a million kilometers per hour and an intense heat that reaches over a billion degrees Celsius.

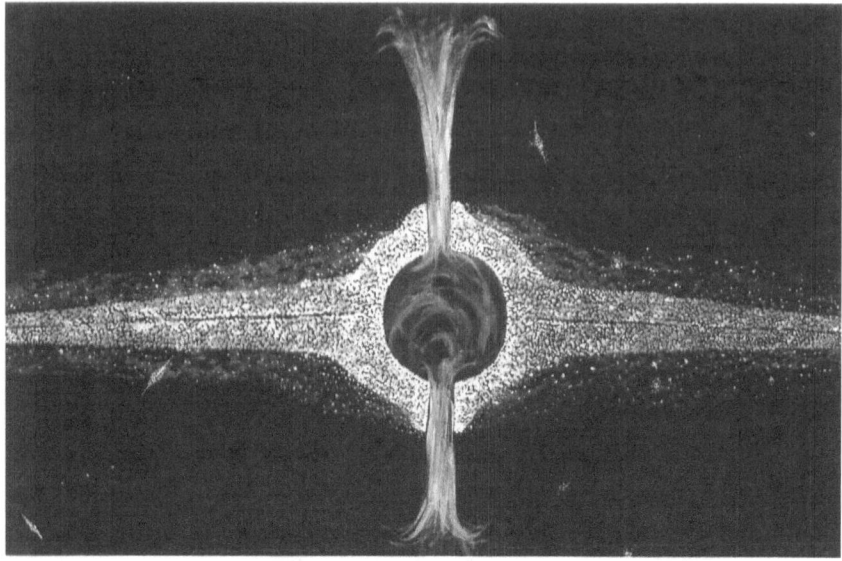

Fig. 8: (Painting of a galaxy sliced in half, showing the total pulverization has been completed. The billions of older stars outside the north and south poles of the galaxy spread apart, hallowing the polarized portals to open without interference. The galactic pressure that has been brewing inside the black hole during the pulverization causes two galactic geysers of very hot subatomic particles to erupt.) By Vincent L. Di Paolo, 2017.

These eruptions most probably last for hundreds if not thousands of earthly years as the two geysers might well reach the length of many galaxies together. Once, all the subatomic particles have been shot out into the cosmos, the billions of old star systems begin to crowd the two polar portals one more time as they will wait for the two gargantuan mouths to open again and begin a new job of recycling a new batch of old star systems.

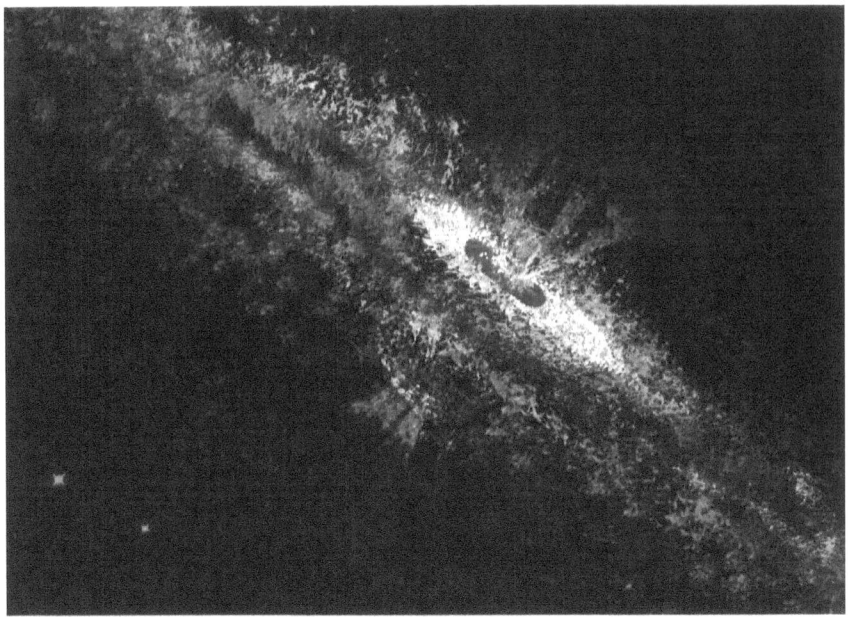

Fig. 9: (Painting of a galaxy showing the polarized geysers are near their end as the spin of the galaxy pulls back all of the subatomic particles towards its suburbs, creating new nebulae or adding to existing nebulae.) By V. L. Di Paolo, 2017.

In Fig.9, thousands of years have passed and the astronomical number of subatomic particles that have been spewed out of both polar mouths of the gargantuan galactic black hole have been pulled back by the powerful spin of the galaxy to join the galactic suburbs of nebulae or to form new ones. Thousands or possibly millions of earthly years will pass as the temperatures of those nebulae decrease, cooled by the cold cosmos around them. However, the temperature of nebulae still remains above freezing *(several degrees Celsius)* when the subatomic particles *(2 upper quarks + 1 lower quark join together to form a proton attracting an electron in its small magnetic field to form hydrogen and 1 upper quark and 2 lower quarks join together to form a neutron)* **begin to reassemble into astronomical number and begin to form an array of molecules** *(molecular hydrogen, water, hydrogen sulfide, ammonia, methane and so many more are present)* **in the ever active nebulae.**

A galactic black hole continues to do job after job for the life of its galaxy, its function so precise in the intake of millions, or possibly billions of old stars and their distribution before beginning disintegrating them *(during each job)*. It is a ***well-oiled super machine***, so careful not to take in too many extra old stars, which might cause a serious perturbation on its spin, which would possibly tear and damage its galaxy. Galactic black holes have been programmed to take in the correct total mass of old giant stars to guarantee a smooth and perfect job in recycling them without causing any perturbations in their equatorial spin. Thus, their respective galaxies are safe and the constant creation of new stars is guaranteed in the nebulae *(stellar nurseries)* formed by the constant astronomical number of subatomic particles spewed out of their respective black holes after each and every disintegrating job.

CHAPTER 5

PROBABILITY OF SUN-LIKE STARS IN OUR UNIVERSE

The simplistic idea that we are the only planet with intelligence greatly offends my thinking and my very existence. Most people on planet earth, convinced by beliefs that have no scientific base at all, truly believe that mankind is the only intelligent group of beings in the universe. The Supreme Matter *(God)* at the very beginning of time *(the First Big Bang)* sent all the DNA and Intelligence flying throughout the young spherical universe, giving it the power to create, recycle and create again, and to expand cosmologically. Therefore, the original DNA and Intelligence *(Supreme Matter)* equally spread throughout the ever expanding cosmos, protected by *alexion* from incredibly high and low temperatures and super destructive winds *(in black holes and their powerful ejections of sub-atomic particles)* during every developmental cosmic stage. Intelligence is everywhere throughout galaxies and universes!

I decided to write the following mathematical formula to illustrate the high probability of stars that are very similar and as healthy as our sun in our universe and the high probability of fertile earth-like planets that exist within it. In Fig. 10, I try to explain as simply as possible the very existing probability of stars and planets that are similar in size and as healthy as our sun and earth.

> **Probability of Sun-like Stars Within Our Universe** (Vincent L. Di Paolo, 2014)
>
> M = mass of the sun = 1.989×10^{30} kg.
> Life is possible if M is in between
> $1.5\bar{9} \times 10^{30}$ kg. and $2.3\bar{9} \times 10^{30}$ kg.,
> also, stars that are within the following range can be very similar to our sun:
> $M \sim 1.989 \times 10^{29}$ kg. $\Longleftrightarrow$ $M \sim 1.989 \times 10^{32}$ kg.
> This probability gives us .20 or 1/5 of all stars within our universe, which contains a minimum of one trillion galaxies, each galaxy averaging 300 billion stars. ∴ Life can possibly exist if M is in between:
> $M \sim 1.5\bar{9} \times 10^{30}$ kg. $\Longleftrightarrow$ $M \sim 2.3\bar{9} \times 10^{30}$ kg.
> or in between
> $M \sim 1.989 \times 10^{29}$ kg. $\Longleftrightarrow$ $M \sim 1.989 \times 10^{32}$ kg.
>
> number of Ms = 1,000,000,000,000 galaxies × 300,000,000,000 × .20
>
> number of Ms = $\dfrac{10^{11} \times 3 \cdot 10^{10}}{5} = \dfrac{3 \cdot 10^{22}}{5}$
>
> = $.6 \times 10^{22}$ stars
>
> = 60,000,000,000,000,000,000,000 stars like the sun
>
> no. of Ms = 60 sextillion stars as healthy as the sun
>
> (even if there is life in only one star system out of every 1000 sun-like stars, that would give use 60 quintillions of star systems with intelligent life within our known universe.)

Fig. 10: (Formula explaining the probability of sun-like stellar quantity in our universe.) Vincent L. Di Paolo, 2014.

As you can see through this basic formula in Fig.10 that our universe contains approximately 60 sextillions *(that is 60 trillion multiplied by a billion)* **healthy stars that are very similar to our sun in size and in production of helium and all other elements needed to**

give life to their fertile planets (*This number is based on the probability that our universe contains about a trillion galaxies. Many scientists believe that our universe may contain several trillion galaxies. That would at least double or triple the amount of healthy solar-like stars and planets with life and intelligence*). **If each one of those stars has a planet like our Earth 60 sextillion planets would have life and a development of intelligence. Even if there would be a fertile planet for every thousand healthy solar-like star, that would still give us 60 quintillion planets with life and developing intelligence** (*that would be 60 billion multiplied by a billion*). **However, most of my students throughout my forty-five years as an educator truly believed that we are the only intelligent beings in the Milky Way; and, most people I know also believe in this minimalist thinking about intelligent life in our galaxy. One fertile planet with intelligence and rich with so many living animals and plants per galaxy would still give us, at the very least, a trillion fertile planets with intelligence and with a great variety of animals and plants throughout our universe. However, it is mathematically most improbable that only one planet with intelligence exists within our Milky Way. Intelligent life within our galaxy is mathematically probable in millions of distant fertile planets orbiting around healthy stars that are very similar to our sun.**

CHAPTER 6

EARLY DEVELOPMENT OF GALACTIC BLACK HOLES

In this small chapter I want to share my vision of how galactic black holes began to develop in today's galaxies at a young stage of their galactic development. By understanding my vision through language and drawings the reader will have a better grasp of future development of galactic black holes in very young galaxies that we are able to discover with today's powerful telescopes. I will use our galaxy, the Milky Way, as an example because we have more literature and sharp photographs of its beautiful star systems and gorgeous nebulae. The Milky Way is nearly fourteen billion years old with the oldest clusters of stars being between eight to ten billion years old. It is safe to state that stars in our galaxy have a life between 8.5 and 10 billion years. Our beautiful and very healthy Sun is approximately 5 billion years old, a middle-aged healthy star with intelligent life *(planet Earth)*, with 4.5 to 5 more billion years of life before it will be recycled inside the Milky Way's galactic black hole.

Let us go back about 5 billion years in the history of our Milky Way, when it was still a young galaxy nearing early maturation *(like a 17 year old boy or girl)*. At that time, its first groups of stars had reached the end of their lives at the center of a young galactic bulge. There might have been hundreds of thousands or millions of the oldest stars which had ballooned into maximum capability, burning all their planetary systems, fusing enough iron to warrant their implosions into the galaxy's first batch of stellar black holes. Their proximity at the center of a youthful Milky Way permitted them to accrete, creating the very beginning of our galaxy's galactic black hole. At least about 200 million stars within the Milky Way have become supernovae, eventually developing into a mature galactic black hole and sending their astronomical number of subatomic particles *(each holding one bit of information)* into its immediate space. Some of those supernovae made up part of our young solar system, forming a young Earth with important intelligent information that is now part of each one of us. We are composed of subatomic particles from millions of recycled star systems that became part of Earth's prehistory in the development of human intelligence.

In Fig. 11, below, I illustrate that union of the first group of stellar black holes that accreted to form the very beginning of our galaxy's galactic black hole *(a period of millions of years)* during *its teen-age years* approximately 5 billion years ago.

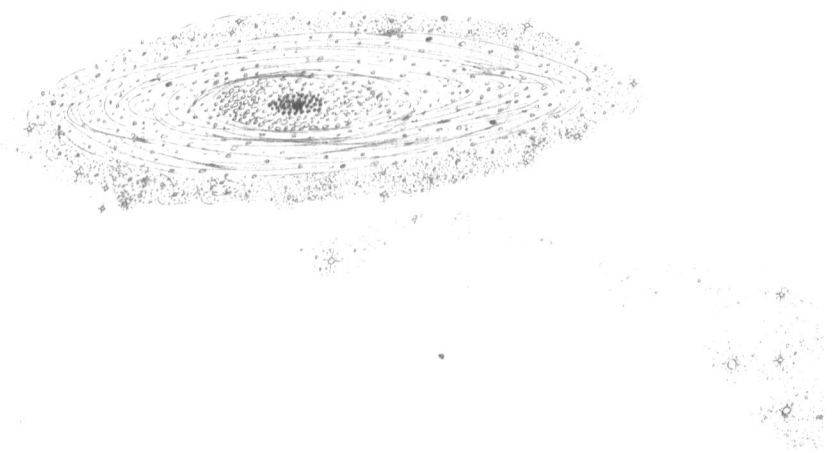

Fig. 11: (A drawing of the first batch of old stars becoming stellar black holes at the center of the galaxy, joining to create the very beginning of the Milky Way's galactic black hole over a short period of millions of years approximately 5 billion years ago.) By V. L. Di Paolo, 2017.

In the past 5 billion years billions of old stars have been constantly recycled by the Milky Way's galactic black hole as it gradually grew with the constant accretion of stellar black holes that were formed in the final stage of larger mega star systems that had reached the center of its young nucleus.

In Fig. 12, you see the constant accretion of a galactic black hole from the millions of stellar black holes forming in its nuclear suburban boundary *(central bulge of the galaxy)*. Finally, after approximately 10 billion years, healthy galaxies develop formidable galactic black holes from the constant accretion of stellar black holes within their nuclear areas. Our galactic black hole cannot be seen as it is surrounded by billions of old star systems *(our Milky Way's central bulge)* that have come to the end of their stellar lives in the past billion years. In the future we might develop possible ways to actually see the formation of a galactic black hole in younger galaxies *(galaxies that are only about 9-10 billion years old)* as their older star systems become stellar black holes and accrete in the nuclei of those galaxies.

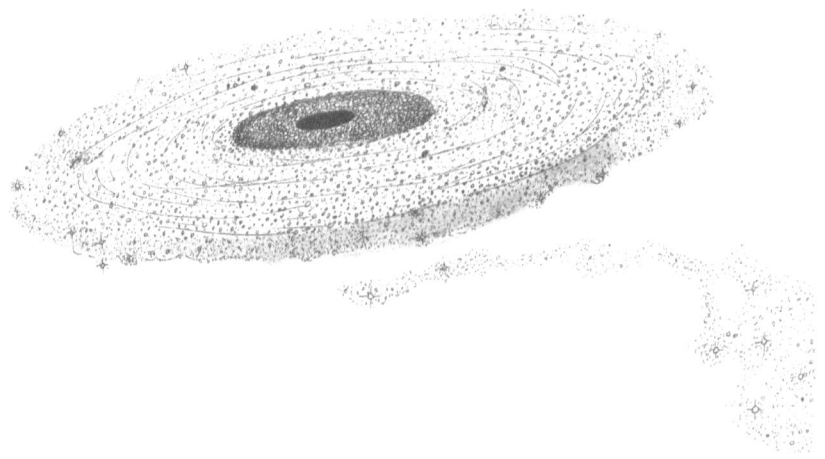

Fig. 12: (A drawing of a maturing galactic black hole in our Milky Way or any other maturing galaxy, showing its accretion of millions and possibly billions of stellar black holes forming at the end of their lives within the central bulge of our galactic super recycling machine.) By Vincent L. Di Paolo, 2017.

I hope that with these two drawings and with my simple explanation you may also envision the formation of galactic black holes from the accretion of newly formed stellar black holes within the center of each and every galaxy. Galactic black holes are super recycling machines that guarantee the continuous pulverization of old star systems and the creation of nebulae *(stellar nurseries)* where new stars are formed from mixed subatomic particles that come directly from all those recycled star systems, keeping healthy and productive developing galaxies.

CHAPTER 7

EARTH: A BEAUTIFUL LIVING PLANET

The Milky Way, our galaxy, is one of a trillion galaxies in our universe. If we could possibly observe our entire universe *(impossible from planet earth)* we might be able to see more than a trillion galaxies that belong to it. The sun, a small to medium star, is one of approximately 300 billion stars that form our galaxy, the Milky Way. The Earth is a small planet of our solar system; but it is definitely the most beautiful one for it is a living planet and our home. I state that we, the people of earth, are so lucky to have been born in one of the most beautiful planets in the Milky Way. There are probably thousands of planets, very similar to ours, in our galaxy and there are most probably, at the very least, trillions of planets in our universe that are very much like our Earth. If we consider mathematical probability, we are so lucky to be living here on planet Earth; and cosmologically, being alive here on this gorgeous planet is like winning the biggest lotto ever. We take it for granted that we live here, not realizing how lucky we are to be alive and living in one of the most beautiful creation of the Supreme Being. Planet Earth received the entire DNA, all the intelligence and the necessary elements to create all types of living plants and animals, which one species finally developed into intelligent beings. It was programmed to do so at the right time, from approximately four and half billion years ago to today and into its future.

Fig.13: (NASA- Telephoto of planet Earth).

Approximately 350 million years ago reptiles developed from some of the fish that inhabited earth's oceans and began to develop an early intelligent brain. The reptilian brain is the most ancient neurological part of the human brain. From the Triassic to Jurassic Periods, some of the reptiles began to develop mammalian qualities as they evolved into cynodont therapsids, laying the foundations for the evolution of mammals. A mammalian brain began to develop, wrapped around its reptilian foundation, neurologically interconnected together. Since the beginning of the Paleocene Period, sixty-five million years ago, primates developed a larger and more advanced mammalian brain. Similar brain development occurred in all other Earth-like planets with Earth-like elements and information.

Approximately five million years ago, towards the end of the Miocene Period and the beginning of the Pliocene Period, early man evolved as the beginning of the cortex wrapped itself around the mammalian brain, neurologically interconnected with both the mammalian and reptilian brains. Modern man's neocortex began to develop approximately sixty to seventy thousand years ago, a small amount of time we call *the missing link*. From hunters and gatherers, early modern man began to use their right hemisphere of their brain, becoming more creative. Early modern man began to paint beautiful and colorful cave wall paintings and began creating sophisticated tools and weapons. First cities were built from fifteen to ten thousand years ago; some of them are under water, around lakes, rivers, and seas *(hundreds of feet below all around the earlier beaches of the Mediterranean and Black Seas, which were then fresh water lakes)*.

In the past two hundred forty years human population has had a negative effect on planet Earth, despoiling all of its natural resources to a critical and dangerous state and polluting the very soil and water systems which have been nourishing us since the beginning of mankind. One hundred thousand years ago, less than one million humans inhabited our planet, living on its purest produce and its fresh spring water. By 20,000 B.C. Earth's human population was approximately between two to three millions, still living on its pristine produce and water. Ten thousand years had passed and mankind's population had doubled to approximately five million and the first cities had been built. Planet Earth was still in pristine conditions, offering its human inhabitants fresh produce and clear spring water. By the beginning of the Roman Empire and throughout Jesus of Nazareth's life, Earth's population had reached approximately fifty to sixty million human inhabitants. Rome, then the biggest city of our planet, had more than a million citizens. Tikal, the largest Mayan city held approximately 500,000 inhabitants; and Teotihuacán was most probably the third largest with approximately 200,000 people. By today's comparison, this planet and its natural resources were still near pristine conditions. However, early evidence of contamination of the soil and water was present: slash-and-burn techniques of farming had negatively affected the soil; and, many of the waters were nonpotable from the blood and dead bodies left in rivers, lakes and wells from battles and human sacrifices.

By the beginning of the Byzantine Empire, the world's population had exceeded 100 million people; and, it might have reached nearly

200 million by 500 A.D. Sometime during the year 535 A.D. Krakatoa, a super volcano on the island of Rakata, located in the Sundra Strait between the islands of Java and Sumatra, erupted with an explosion which equaled about two billion atomic bombs. The explosion was felt around the globe and it was the greatest eruption ever witnessed in the recorded history of planet earth. Volcanic ashes shot into the atmosphere reaching 80 kilometers high. Strong winds created a vast belt of dark volcanic ashes around the Earth, blocking the sun: it was the beginning of the Dark Ages. A grey nuclear winter blanketed most of our planet, covering the rich green landscape of most of its land, and poisoning most of its water systems. Most probably, only the land and waters inside the Arctic and Antarctic circles still received direct sunlight. The large belt of volcanic ashes covered 70 to 80 percent of the earth, blocking direct sunlight. The days were very similar to late dusk and the nights were pitch-black. The moon and the stars were not visible at night, blocked by the thick dark grey belt of volcanic ash. Around the globe, there were no crops for many years. Plants' rings that have been studied clearly show no growth between 535 and 555 A.D. During those twenty years the plague and famine decimated much of the population as it circled this planet several times. Fruits and vegetables did not mature and people got sick and died from eating them. Animals, very much like people, suffered many sicknesses and died young. By 555 A.D. the population had dropped to approximately 50 million, many of the surviving people were weak and sick. During the Dark Ages, the human brain suffered greatly due to lack of good nutrition as it took a big step back in its evolutionary development.

It took 700 years for the Earth to rejuvenate itself; however, its population had only reached approximately 55 to 60 millions by the 12th century. Finally, the Renaissance *(il Rinascimento)* had begun in Italy by the late 1300s due to its rich Mediterranean diet; and, the arts flourished once again after nearly a millennium. By the late 1500s the Renaissance had spread throughout Europe, bringing a rebirth of the arts, writing, music and learning. By this time Earth's population was only approximately 100 millions. The Italian Renaissance had propelled a rebirth in every sense of the word: better food, thirst for knowledge, and a vast development of the arts *(drawing, painting, sculpture, writing, music, drama, opera and finally classical music)*. Once again, the human brain continued its healthy development.

By 1804 A.D., Earth's population had reached its first billion; and, the Industrial Revolution was well in its way, burning a great amount of coal and smelting huge amount of iron ore. Most of the major cities throughout the world had become extremely polluted by early 1900s. With the inventions of machinery, trains, cars, busses, trucks, airplanes and large ships, coal and oil were heavily consumed. By 1960 A.D. Earth's population had reached three billions. Today's population of 7.75 billions has more than doubled in only 55 years. In the next eight years, by 2025 A.D., this planet's population will reach over nine billion inhabitants.

Global warming has naturally occurred many times on our planet for it is a natural reoccurring period between ice ages. At the end of each ice age warming periods we call interglacial periods, lasting from fifty to one hundred thousand years, warmed up this planet while melting a large percentage of the north and south ice caps. During interglacial periods the sea level rose several hundred feet, flooding many fertile river valleys inhabited by early man. Approximately ten thousand years ago a new global warming commenced with the melting of the Fourth Ice Age which had begun approximately twenty-five thousand years ago. Is it possible that the two to three million people living during this last ice age shortened it by all the continuous fires they diligently kept feeding to keep warm, triggering an early beginning to today's interglacial period *(global warming)*? As the population of our planet increased exponentially a lot more carbon dioxide was produced. Also, all the volcanic explosions in the past ten thousand years have sped up the rise of today's global warming. Earth's population has definitely helped to speed up the release of carbon dioxide into its atmosphere.

"The Earth, our home, is beginning to look more and more like an immense pile of filth...Climate change is a global problem with grave implications. It represents one of the principal challenges facing humanity in our day." wrote Pope Francis in his latest encyclical, addressing a serious and impending catastrophe that we all have created as a world population.

I admire Pope Francis very much for he is the first pope to ever tell it like it is. I do hope that he realizes that the biggest culprit is the ever growing population, which keeps increasing exponentially. The Catholic Church, India, and other countries *(with serious population growth)* must impose an immediate population control. We do not have the luxury of time to think about this critical problem. Today's population of 7.75 billion people uses more than one billion vehicles; and, by 2025, which is the very near future, a population of more than 9 billion people will be using more than two billion vehicles, nearly doubling today's emission of carbon dioxide. We cannot deny that with today's total vehicles, factories, buildings, houses and other machines we have definitely greatly increased the amount of carbon released into our atmosphere, speeding up earth's natural global warming.

Millions of unsold new cars are parked in hundreds, maybe thousands of gigantic parking lots, left to rot as their engine oil filters down; and, car manufacturers keep on producing millions of new cars. All the unsold new cars will be parked in newly created parking lots as next year's models are being manufactured. All this mechanical waste does not include billions of older vehicles in junkyards found everywhere on this planet. How much longer can we keep polluting the landscapes of our beautiful planet? Our oceans have become so polluted that there are seas of floating garbage everywhere in all of earth's water systems *(oceans, seas, lakes, rivers and creeks)*. The acidity of the oceans has reached a critical stage, leaving us a small window of time to clean it up. If we do not clean up the enormous amount of garbage we have created on both land and waters within the next ten years it will be too late as we reach the point of no return when this planet's population will surpass the nine billion count. You, the reader might ask if it is already too late. Surely, if we do not slow down or arrest the population growth within the next ten years it will be, then, too late. We have a small window of about ten years to stop population growth; and, come up with the money and manpower to clean up all the garbage we have produced on land and waters, creating millions of jobs around the world and recycling all of that garbage into new usable products: paper, plastics, wood and other products to be used in creating new conglomerates.

In the past ten years, the U.S.A alone has averaged a yearly expenditure of approximately more than 700 billion dollars for defense. Russia has more than quintupled its defense budget because Russian leaders feel the built-up NATO lines of defense near the Russian borders. China and many other Asian and Middle Eastern countries have also augmented their defense budgets. Yearly, trillions of dollars are spent on new jet fighters, new warships, submarines, new tanks, army trucks, jeeps and weapons. If we take only ten percent of that money spent in defense and use it to create jobs in cleaning up the garbage found on land and water, each country would still have ninety percent of their original defense budget, which would be more than sufficient. That would mean about seventy billion dollars just in the U.S.A., which would create nearly two million jobs in cleaning the U.S.A. and all of its water systems. If the European countries, Russia, China and all the Asian countries, and all the other countries on earth would do the same by creating new jobs to clean up our planet, we could possibly do it within five years.

From battleship carriers to tankers, from cruise liners to commercial fishing boats, from ferries to private fishing boats, we have millions of boats crossing our oceans, seas, lakes and rivers. If we would pay a specific amount of money per ton of floating garbage in Earth's water to any boat that brings it in to be recycled we can clean our oceans, seas, lakes and rivers within five to ten years. As citizens of this most beautiful planet *(especially leaders of all countries and presidents of corporations and companies)* we must resolve this alarming problem as soon as possible. A world committee must be created immediately, funded by all nations, beginning with the U.S.A., creating millions of jobs around the world in the total cleaning up of this planet. Corporations and companies must also give ten percent of their net profits to help this formidable cause. If we choose to ignore our population explosion and the immediate need of cleaning up this gorgeous planet, it will be too late in a few years, when the population and the garbage will reach an insurmountable state.

Since 2010, throughout the world, a yearly average of approximately 1.5 billion tons of food is thrown away as garbage while nearly 3 billion people are starving. All that discarded food equals 3 trillion pounds of food wasted yearly. If you divide 3 trillion pounds by 3 billion people it would give one thousand pounds of food per starving person each and every year. Almost 3 pounds of food daily per person *(3 billion starving people)*. That would solve world hunger and we would have less garbage each and every year. The biggest culprits of food waste are the U.S.A, European countries, and several other rich countries throughout the world. The food industries choose to throw away whatever they don't sell rather than give it away to the poor people in every major city, towns, and villages. We have become so addicted to money that we would throw away whatever we cannot sell *(food, cars, etc...)* instead of helping our unfortunate fellowmen. By being so selfish we choose to ignore the starving poor and create more waste that will soon be nearly impossible to clean up. All in the name of money we have fleeced all the resources of our planet and have given back mostly garbage and pollution. The leaders of every country, the presidents of major corporations and every other smaller businesses, and each one of us need to try to redeem ourselves by cleaning up the mess we have created so that we could guarantee our children and grandchildren a chance to live in a cleaner and healthier planet Earth. If there is a little decency left in each one of us we must embark on the cleansing of our planet immediately. We cannot wait any longer for we have reached the point of no return.

CHAPTER 8

THE HUMAN BRAIN: A COSMIC NEUROLOGICAL DEVELOPMENT AND A MICROSCOPIC REFLECTION OF THE UNIVERSE

The human brain directly evolved from planet Earth, a product of the solar system and the Milky Way, one of a trillion galaxies of our universe, which is a direct product of the very First Big Bang, created by the Supreme Matter. Planet Earth is our mother, a fertile and living planet. It took approximately four billion years for the Earth to produce the first brain, the reptilian brain. During those four billion years a similar development of the reptilian brain evolved on quadrillions, possibly quintillions of Earth-like planets within the trillion galaxies of our universe. These last four billion years of evolution are not the first batch of Earth-like planets producing a reptilian brain. We are at least the second batch and possibly the third, whose information had come from previously recycled system of stars throughout all of the galaxies. It is highly probable that a similar reptilian brain developed simultaneously on several quintillions of Earth-like planets of our universe, some a little earlier or a bit later than four billion years. All of the Earth-like planets were all formed from the same elements, information *(intelligence)* and DNA that were released from the same Big Bang, thus sharing all probabilities. In the past one hundred years, our human brain has gone through a major development, especially in our scientific thinking. It is unthinkable that such a huge percentage of the population still believes that the Supreme Being chose only planet Earth, among quintillions of fertile planets in the cosmos *(only in our universe)* to have intelligent beings. This type of primitive thinking is not acceptable today. "What about Jesus Christ?" you might ask. As I said in the preface of this book, Jesus is my first hero and best human being ever, who taught us that we are all children of The Supreme Being and showed us that through love we will conquer everything and make Earth a great and wonderful planet. I am afraid that we are still having a lot of trouble following his noble and loving teachings.

As we are a product of the cosmos, directly programmed to evolve into intelligent beings by the Supreme Being, so are the quintillions of fertile planets with equal or more intelligent beings within our universe *(excluding all other universes)*. As there are trillions and trillions of galaxies out there in the cosmos, so there are quintillions of intelligent planets at the very least. Does that make you feel small? It does make me feel small, finite and not very significant; but simultaneously, it makes me feel great, important and bright to see the whole picture in my mind.

More than three hundred million years have passed from the Cambrian evolutionary explosion of a multitude of animals on planet Earth before the first mammals and the first dinosaurs began their own evolutions. Approximately two hundred twenty million years ago, during the end of the third quarter and the beginning of the last quarter of the Triassic Period the first mammals developed simultaneously with the first dinosaurs. It was during the end of the Triassic Period that a new brain began its long evolutionary growth, wrapped around the very important neurological base we call the reptilian complex. That new brain was the beginning of the mammalian brain, neurologically connected with its reptilian complex, which was the base upon which the new brain evolved. Both the reptilian complex and the new mammalian brain worked together for more than two hundred million years in developing a full mammalian brain. Thus all mammals shared that development; but, some mammals' neurological system evolved quicker than lesser mammals. The mammalian brain might not have evolved in every fertile Earth-like planet; in some planets *(with oceans covering their entire surfaces)* it might have been replaced by a second reptilian brain wrapped around the reptilian complex, possessing similar features of the mammalian brain.

By the Tertiary Period of the Cenozoic Era the mammalian brain had greatly evolved, producing *hormones* for growth, *myelin* for the insulation of the axons of all neurons, and *endorphins* to fight sicknesses that would attack all mammals. This mammalian evolution most probably happened to a large percentage of all Earth-like planets; and, in some of those planets that neurological evolution possibly might have happened with their dinosaur families that might have evolved as mammals or developed a second reptilian brain.

It was at the very end of the Tertiary Period, approximately three million eight hundred thousand years ago that the neocortex began its evolutionary growth, neurologically connected to both the mammalian brain and the reptilian complex. However, the largest and most impressive growth of the neocortex took place only about sixty-two thousand years ago as modern humans distinguished themselves from *homo neanderthalensis*. This sudden evolution is sometimes referred to as *the missing link*, which I will not discuss it in this book: it will be addressed in another book that I am writing. The quick development of the neocortex in the past sixty-two thousand years accelerated humans' neurological evolution. However, the human brain greatly suffered a drawback during the Dark Ages and most of the Middle Ages. We would have been much more advanced if the Dark Ages would have never occurred. I firmly believe that the development of the neocortex in humans is only the beginning of a larger galactic and eventually a super cosmic neocortex, which will continue to evolve if we will not annihilate ourselves or completely destroy this planet. The larger galactic neocortex most probably exists in a significant percentage of intelligent humanoids living in other planets of the Milky Way and other galaxies. A smaller percentage of intelligent beings in other living planets throughout our universe might have reached early development of the cosmic neocortex, the ultimate development of the cosmic brain.

At birth, babies are born with a triune brain and a young

cerebellum, with a minimal myelin insulation of the axons. However, babies are born with an average of more than one hundred billion neurons and more than fifty trillion neuroglia cells *(also known as **glial cells**, which produce myelin and defend the brain from microbes)* that they will ever need as adults, each neuron connecting with hundreds of other neurons by thousands of dendrites. The human brain is a young microscopic reflection of our galaxy; and, all other cosmic brains throughout the universe are born to baby humanoids that evolve like humans or evolve in similar ways. A newborn's brain cannot do much except survive as a living animal. It will take time *(up to 18-20 years)* before the human brain is fully developed and insulated to be fully functional.

Paul MacLean, while working at the Laboratory of Brain Evolution and Behavior, at the National Institute of Mental Health in Washington, D.C., discovered that the human brain processes information through three distinct yet neurologically interconnected sub-brains: **the Reptilian Complex** *(the primal mind and the brain's receptionist, as it evaluates all incoming messages and decides to send them to specific parts of the brain or to simply hold them);* **the Limbic System or Mammalian Brain** *(it is our emotional and nurturing brain, our motherly brain and the brain's thermostat: it regulates hormonal discharges to maintain homeostasis and mediates emotional and social behavior; it manufactures endorphins to regulate pain and fight influenza and other sicknesses, hormones for growth, and myelin to insulate each and every neuron; the Hypothalamus and the Pituitary Gland are two very important parts of the Limbic System that begin to coordinate emotional behaviors and activate the body's stress mechanisms resulting in a series of hormonal and muscular reactions);* **and the Neocortex** *(our rational and metaphoric brain which is divided into a left and a right hemispheres, as it takes all sensory data and interprets it and takes immediate actions based on problem-solving techniques and models that it has incorporated into its system; it also controls conscious muscular movements, including speech, reading, writing and all of the arts).*

During the first five weeks of the embryonic development, the neurological growth looks very much like the brain of a reptile *(the reptilian complex)*. By the fifth and sixth months *the mammalian brain* has grown around *the reptilian complex*, totally neurologically interconnected with it. It is during the last three months that the *neocortex* grows around the mammalian brain, completely interconnected with both the mammalian and reptilian brains. This neurological development occurs in all prenatal babies of every living planet with intelligent life throughout the entire cosmos. However, if some of the intelligent beings of other living planets have gone through more advanced neurological developments, their brain during prenatal growth will develop a second *(galactic brain)* and possible a third *(cosmic brain)* neocortex, stretching the humanoid skull at its softest point. In the future, if humans still manage to survive, their brain development will increase, at least by two more neocortices *(the galactic neocortex and the cosmic neocortex)*. These two extra neurological developments might require a longer prenatal period *(possibly ten to eleven months)*. See Fig. 14 below, shows the neurological development of the Triune Brain during pregnancy.

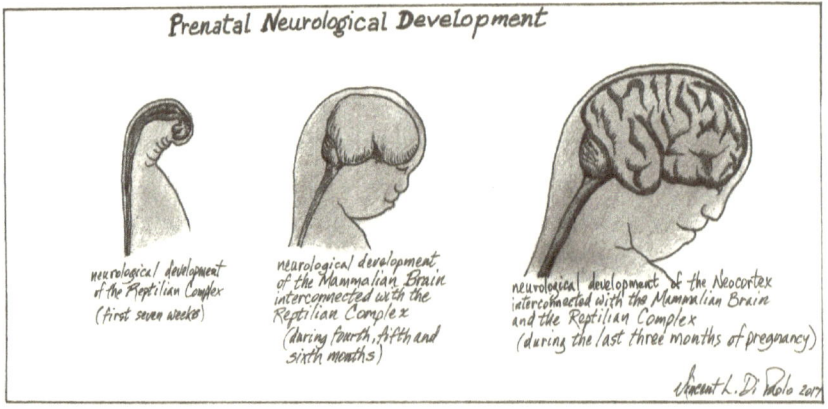

Fig.14: The development of the Triune Brain during pregnancy (drawing by Vincent L. Di Paolo, January, 2017).

All women should cleanse their bodies of too much Omega-6 fats that might be present within their systems at least six months before conception by changing their diets into a rich Omega-3 diet with a generous dosage of antioxidants to guarantee a healthier neurological development of their future babies. Simultaneously, a mother-to-be should stop smoking and drinking alcohol or taking drugs at least one year before conception. A healthy nutrition *(rich in vitamin Bs)* and lots of love from husbands, relatives and friends will play a major role in the physical and neurological developments of their babies' future embryonic and fetal stages. The use of alcohol, tobacco, marijuana and other drugs will have a negative effect on both the physical and neurological developments on prenatal babies. Also, too much fatty meat *(Omega-6)* will cause problems for the developing embryonic brain. We can get healthier Omega-6 oils from linoleic acid *(oils from evening primrose, black currant seeds, sunflower, safflower, soy, sesame and blue-green algae)* rather than from animal fat. A good balance of Omega-6 and Omega-3 oils, coupled with antioxidants and all of the vitamins *(especially B-complex)* are vital for a healthy neurological development, free from neurological disorders. Excellent classical and softer music will positively stimulate the developing brain. For mothers, pregnancy should be accompanied by love, care and excellent nutrition. Most vitamins, especially the B-complex, directly affect the foundation of the neurological system and the developing brain during the entire pregnancy. The lack of the vitamin Bs can cause babies to be born with auditory dysfunctions, neurological disorders and possibly autism. The B-vitamins are called the brain vitamins and are essential for a healthy brain development during the entire nine months. Once the baby is born the B-vitamins continue to be vital for a healthy brain development, especially during the four stages of myelination. Choline, lecithin, vitamins C, B1, B5 and B6 are needed to synthesize the neurotransmitter *acetylcholine.* To synthesize the neurotransmitters *dopamine, norepinephrine* and *epinephrine* the brain needs L-phenylalanine and L-tyrosine *(two amino acids)*, vitamins C, B3, B6, B12 and copper. The neurotransmitter *serotonin* is synthesized with foods rich in carbohydrates *(vegetables, breads and pasta)*, L-tyrosine, L-tryptophan and vitamin B6. All other neurotransmitters also need a variety of vitamins, especially B-vitamins to be synthesized. Also, neurotransmitters use B-vitamins as fuel for traveling throughout the vast neurological system *(approximately 260,000 miles of interconnected neurons and glial cells)*. As adults we need the B-vitamins to continue a healthy and super-fast thinking brain. I will give more details on the importance of B-vitamins when I will illustrate the important stages of myelination and production of neurotransmitters.

The birth of a child should be followed by constant care by the parents and the best nutrition, mothers' milk. Women should be given at least a full year or more of paid maternity leave and fathers should be given at least three months of paid paternal leave to help a very demanding task of taking care and nurturing their newborn.

Reading to the baby from the very first day is as important as their mothers' milk. Even though a neonate might not understand the words being read he or she recognizes the parents' voices, stimulating early acquisition of the foundation of their mother language by sound, smell, vision and touch. Reading from the first day is so vital to the development of the foundation of that language. After several months of their mothers' milk and daily listening to reading of early children's books babies begin to babble, practicing their first sounds *(syllabic nasal consonants such as KAH! GAH! are produced at the back of the mouth where the tongue meets the palate; and, by six to nine months babbling and reduplicating babbling such as "NA-NA-NA!" and "DA-DA-DA!" are frequent and universal)* and expanding their neurons' dendrites through *arborization* and the foundations of memory banks throughout their very young brains. All this babbling is the necessary exercise for babies to begin uttering their first words. At this stage, children's ability to comprehend their mother language precedes and exceeds their ability to speak it: they know much more than what we think they know and their memory are incredibly exact, vivid and meaningful. So, be careful how you speak in front of your babies or toddlers. Parents should keep reading to their children so that the neurotransmitter *dopamine* gets hooked on books rather than electronic gadgets. I am horrified when I see mothers and fathers giving their young children electronic games, as they ride on carts, while they shop: in these cases *dopamine* gets hooked on gadgets and loses interest in books. Later, those children have a difficult time with books and at school because their *dopamine* has found pleasure in electronic games and have become addicted to them. During early elementary education, teachers and parents will have a tougher time to instill on those children the importance of speaking, reading, writing, mathematics and studying.

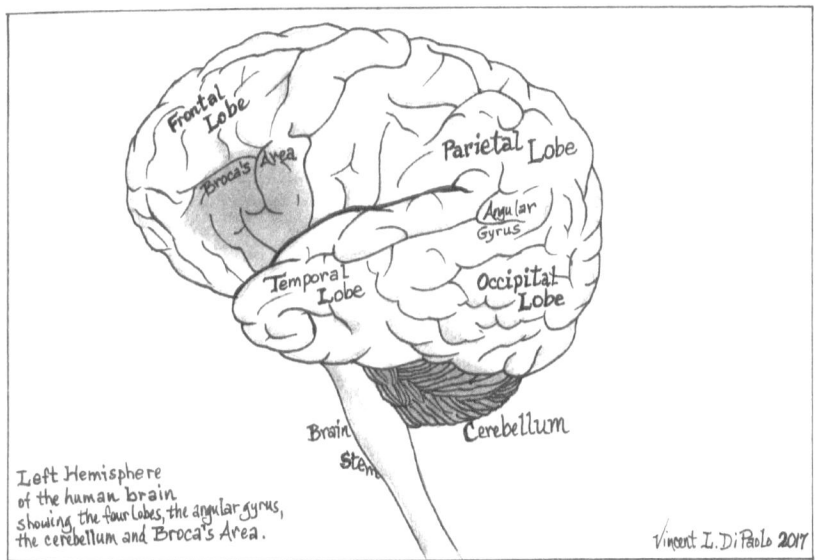

Fig. 15: Left Hemisphere of the human brain, showing the cerebellum and Broca's Area. Drawing by Vincent L. Di Paolo, 2017.

The Neocortex, which is divided into the Left and Right Hemispheres, is our rational and metaphoric brain. All information that comes into our brain through our senses *(with exception of the olfactory system: smell)* goes directly to the *thalamus,* which interprets all that data and sends it to the appropriate parts of the neocortex. The thalamus is the interpreter of all the information *(except smell)* we absorb by our senses. The Neocortex's actions are based on problem-solving techniques *(a right hemispherical solution)* it has incorporated into its system and synthesized in a personal thought or decision. The conscious control of muscles, including speech, reading, writing, mathematical problem-solving and all artistic actions reside in the Neocortex. Both left and right hemispheres are divided into four lobes: the Frontal Lobe makes the final decisions and sends them to the *cerebellum* from the information directly processed by the Parietal Lobe, which immediately gives the information to the Frontal Lobe; the Occipital Lobe processes vision and the Temporal Lobe processes sound and smell. Each thought and decision *(made after synthesizing related information)* is processed in a left-to-right and back-to-front organizations. If the Frontal lobe decides to practice tai-chi the Cerebellum *(the muscle coordinator)* gives you specific muscle commands to all the detailed steps and movements. The Neocortex organization is from left to right hemispheres, connected by the Corpus Callosum, a bridge made up of approximately 200 million nerve fibers. Each hemisphere operates differently from the other and each one processes different cognitive functions. The left hemisphere *(the super secretary)* is concerned with detailed information and it formulates rational, logical, and abstract thinking *(following its innate logical rules of grammar and mathematics)*. It also processes activities in a sequential order, like language *(speaking, reading and writing)*, arithmetic and problem solving *(where it needs the synthesizing ability of the right hemisphere)*. We use our left hemisphere in operating today's technological tools and gadgets.

The right hemisphere operates in a holistic way; and, for that reason it is called *the metaphoric mind* because it focuses on the greater picture rather than the details. It sees the entire cosmos rather than the individual star, planet or moon. It sees the entire beach rather than concentrating on each grain of sand. Its intuitive thinking accelerates it to a fast conclusion without going through the left hemisphere's logical and analytical steps. Its holistic and speedy thinking usually results in inventive and creative acts. Fine arts, music, drama, dancing, inventions and creative stories come from our right hemisphere.

Broca's Area, found in the lower part of the frontal lobe within the left hemisphere is where the foundation of the mother language develops. Speaking, reading, writing, grammar, drawing, penmanship, mathematics, second languages and all other language-related subjects like history, geography and science develop around it. However, the best way of learning is using the synergic powers of both left and right hemispheres. Education should give children the chance to conceptualize their ideas and show them the skills to work out all the needed details. The best curriculum contains a balance of both ideas and words, wholes and sequences, synthesis and analysis, space and time *(which must include a daily positive experience of the arts and physical education)*, synchronized with each of the four myelin productions.

CHAPTER 9
MYELIN PERMITS HIGHER INTELLIGENCE

The Limbic System or the Mammalian Brain produces growth hormones and secretes other hormones from the endocrine glands for physical development, controls the production of endorphins to fight microbes and most sicknesses *(depends on proper nutrition)*, and stimulates myelin production by oligodendrocytes in the central nervous system for insulation of the axons of all neurons *(permitting more electrical impulses, quicker messaging, more information, and quicker memory retrieval)*. Since early 1800s most scientists and educators believed that the human brain progressively grew during the first 16-17 years of life. Today's education is still based on that archaic nineteen century theory, which both the American and European educational systems are based on. The idea that the human brain progressively grew during the first 16-17 years of life was based on an old Prussian psychological and educational theory. Even though today's education has taken bold steps in learning it is still based on a progressive brain growth *(from birth to high school)*. Thanks to the great neuroscientist and neurosurgeon, Herman T. Epstein, a new light was shed in the 1970s on how the human brain grows. After a few thousand brain operations on children of various ages Epstein clearly witnessed that the brain does not progressively grow until age 17; however, its growth is directly associated with four myelin productions and plateaus *(a stop of myelin production)* in between. H. T. Epstein, Paul MacLean, Richard Restak, and many other neuroscientists clearly showed us that the last pound of the human brain develops in four myelin productions.

Learning of new materials and subjects are directly associated with each of the four myelin productions in the Limbic System

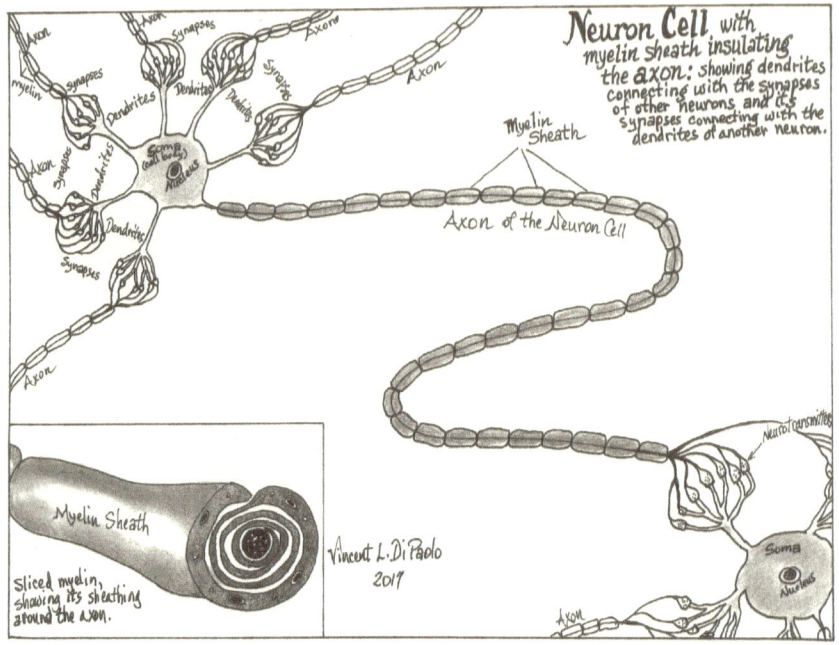

Fig.16: Myelination of all axons of all neuron cells, showing a myelinated neuron connected with other neurons. Myelin insulates all axons, permitting quicker messaging. Drawing by Vincent L. Di Paolo, 2017.

or *Mammalian Brain*. Also there are very meaningful differentiation of myelin production between boys and girls. This differentiation of myelin production might not be universal within the cosmos; although there might well exist intelligent planets with the same myelin production differentiation *(based on the bell curve)*. Therefore, we can safely say that myelin permits or allows learning and the expansion of superior intelligence. Without myelin or with very little myelin humans would be no more intelligent than lower mammals.

As we evolve, the production of myelin is greater because today's scientists are improving our daily diets with the proper nutrition and vitamins needed by the Limbic System and the rest of the brain and its neurological system throughout the entire body. The production control of myelin by the Limbic System is directly associated with the production of hormones *(for physical growth)*. The difference of hormonal productions between boys and girls directly affects a differentiation of myelin production between them at each stage of the four myelin productions or brain spurts. The first two myelin productions occur only in the left hemisphere of the brain, which deals with the acquisition of the mother language, second languages, speaking, reading, writing and the four mathematical foundations *(directly affecting the learning of all elementary subjects)*. The third and fourth myelin productions occur later in the right hemisphere of the brain. After the first two myelinations the human brain is imbalanced resulting in various behavior problems. It is only after the end of the final myelination that the brain is balanced and young adults are able to fully benefit from a totally insulated brain and neurological system. In the next two chapters I will discuss in detail the importance of excellent nutrition during every myelination and how it affects learning, intelligence and memory during each brain growth; and, the immediate need for educational systems to synchronize learning of new and more difficult materials with each of the four stages of myelination, creating new and extremely effective curricula.

CHAPTER 10
MYELINATIONS I & II: INSULATION OF THE LEFT HEMISPHERE

When a baby is born his or her brain weighs approximately one pound; but it contains all the neurons needed for adulthood. During the first ten to eleven months of newborn babies their brains double in weight; but 95% of that second pound is the growth and neurological development of the *cerebellum* (*the automatic pilot*), which controls the physical movement and coordination of the muscles and skeletal system *(in conscious and unconscious actions)*. When babies are loved, cared for and given excellent nutrition and daily reading, it stimulates a production of new dendrites and synapses in their developing brains. Young babies begin to move their arms and legs upwards because their cerebella are developing. A hanging play station stimulates the movement of the limbs, bringing a certain satisfaction and accomplishment to young babies; they smile and gurgle as they feel good to touch the hanging objects. However, babies' development can be reduced and slowed down if there is abuse, hatred and malnutrition in their lives. Unfortunately, the babies that get no attention, very little stimulation and poor nutrition by their abusive parents will have a lesser myelin production. Those types of parents should never have had children! All young people that are thinking of getting married and eventually having children should educate themselves on positive loving parenting.

During the next six to seven months, babies begin to turn by using their arms and legs. They begin to do early push-ups, raising their torsos; and, they begin to crawl, stand up and take their first steps after many falls. By the end of their first year, most babies are walking and some are already running. The production of new dendrites and synapses quickly grows, making billions of connections throughout the entire neurological system. A mild coating of myelin occurs throughout the neurological system, especially in the cerebellum, from six to twelve months, enough to permit the neurotransmitters to travel at an easy pace. By the end of the first ten to twelve months babies' brains have doubled in weight because of the quick development of the cerebellum. The third pound of a young adult brain will consist mostly of myelin insulation of the axons of all neurons. The third and final pound of the brain will grow during four stages of myelination *(from 18 months to 18.5 years old)*.

Excellent nutrition, constant attention and daily reading to your baby help to establish a solid foundation needed for a strong mother language acquisition, which is vital as the base of learning. An excellent diet also guarantees a great production of more than 100 different neurotransmitters *(electrochemical messengers)* each with a specific specialization in helping the human brain in communicating with every function of learning, memory, intelligence, physical growth, motivation, focus, emotions, perseverance, energy, coping with daily stresses, sleep, sexuality and every other aspect of modern human neurological development. There may possibly exist several hundreds more neurotransmitters operating within the human neurological system; those neurotransmitters will be discovered and named as the human brain and body keeps developing at an ever increasing rate. As humans, we still have a long way to go in our potential future neurological development for we are still at an early stage of the final cosmic brain. I firmly believe that myelination occurs in every intelligent being *(including animals like dolphins and whales)* throughout each and every galaxy of the entire cosmos.

The *first myelination* occurs in the left hemisphere of the brain *(with a great concentration in Wernicke's area and later in Broca's area)* from approximately 18-20 month old to approximately 4.5 years old in girls and from 22-24 month old to 5 years old in boys. There are the possibilities that a small percentage of boys might have an earlier myelination very much like the girls and a small percentage of girls might have a later myelination like the boys. However, the majority of the girls will have an earlier myelination *(approximately 5-6 months before)* than the majority of the boys. A production of approximately 110 grams of myelin will be produced during these three years that will permit the acquisition of the mother language in each and every child. A healthy and rich production of myelin *(105-114 grams)* will occur if there is a healthy and rich diet, rich in vitamin Bs, coupled by stimulation through daily reading of stories, adult-like vocabulary *(no baby talk)* spoken to the child and a lot of love and care by everyone in contact with each and every child.

Unfortunately, that rich nutrition and stimulation does not take place in every family. There is a large percentage of families throughout this world that do not have a rich diet nor a rich family involvement in their children's first learning of their language. In reality, a high percentage of families are starving and there is little or no stimulation for their children. The myelin production in very poor families might be minimal *(40-80 grams)*. Not everyone is born in a wealthy, caring family; not everyone is born equal. This is where we need to work as a society, to guarantee that everyone gets a proper daily nutrition and care. We have so much to do to achieve equality at birth.

During the first myelination *(the foundation of the mother language)* children learn how to talk in their native tongues. If there is a rich myelination during these three years, coupled with good genes, children learn to speak; and, by the end of this period they should be speaking by using simple, compound and complex sentences. Parents know that when their four and a half years old daughter or their five year old son speaks in clear complex sentences they are ready for school. If a five year old speaks in one word sentence *("Drink!"* or *"Thirsty!")* that child is in trouble and will not be ready for school. A child that is ready will most probably say *"May I have a drink, mommy; I'm really thirsty!"* and he/she has developed a solid foundation of the mother language. With a solid *first myelination* children are ready for school and with the second myelination *(in the left hemisphere)* they will be ready not only to learn how to read and write but they will be able to begin their foundation of mathematics and the acquisition of other languages.

After the *first myelination* there is a *plateau* *(a rest or temporary stop of the production of myelin)*. **During the first plateau** *(from 4.5 to 6.0 years old in most girls, and from 5.0 to 6.5 years old in most boys; a few boys might be closer to the girls' plateau and a few girls might be closer with the boys')* **pre-kindergarten and kindergarten teachers should not try to teach more difficult or advanced subjects. Those teachers should be concentrating on fun activities with their young students: dramatic reading; reading interesting stories using puppets; storytelling by teachers acting their parts and with costumes if possible; learning how to count; learning how to speak to others; dancing; learning how to paint; drawing or doodling, and so many more fun activities around their mother language** *(English in the U.S.A.)*. **This is not the proper time to introduce a second language** *(Spanish, French, Italian, German, Japanese and other second languages should be taught during first and second grade)* **or to begin teaching multiplication or division; doing so during the first plateau will not be beneficial to those young students and it might confuse them or turn them off. The use of computers, calculators, I-phones and tablets or technological games should be outlawed at such a tender age when the young brain is beginning to develop a solid foundation and learning styles. Any principal or teachers who insist on using technology with young students** *(from birth to eight years old)* **are slowing down or altering children's neurological development. <u>The use of technology during the first 8 years of young children's lives will alter and possibly even atrophy parts of the young developing brain!</u>**

Most schools in Asia forbid the use of technology in elementary schools, concentrating on developing excellent memories by reading, speaking, writing, mathematics, calligraphy, the arts and physical education. The use of calculators and computers are introduced in secondary schools to facilitate advanced mathematics and science. Unfortunately, in the western world computers are being pushed by companies and adopted by principals of elementary schools and by teachers and parents who know very little or nothing about brain development, especially at such a tender age.

The *second myelination* begins at age six with most girls and it ends by the completion of their eighth birthday. With the boys, the *second myelination* begins at 6.5 years of age for most boys and it ends by 8.5 years old. Again, there are approximately 5-6 month difference with the beginning and ending of the second myelination between girls and boys. Once again, there is a production of approximately 105-114 grams of myelin being distributed around the axons of all the neurons of the left hemisphere during these two years. A smart first grade teacher can tell who is deficient in spoken language and can create a great learning environment for those children by grouping them with children who are speaking at a more advanced level. Also, a healthy breakfast and lunch *(rich in vitamin Bs)* should be provided daily by all schools to children from low income families; many children go to school without eating breakfast and might not have a decent lunch or nothing at all. Free breakfast and lunch to children of low income families guarantees a better production of myelin and learning for those children, especially if those meals are rich in vitamin Bs.

When I was working on my master at McGill University I did several studies with young first and second graders *(approximately 60 students)* who were deficient in language and most of them spoke in one or two words sentences. I used Noah Chomsky's ***Dramatic repeated readings***, recorded at three different speeds. I would tape a passage of a play at a slow speed, allowing those students to understand and read each and every word following the slow dramatic reading. I even added sound effects to the tape to dramatize the passage. Those students would repeat that passage as they followed the dramatic taping five to six times. By the fifth or sixth practice, they had mastered that slow fluency in their reading, enunciating each and every word. I would prepare a second tape of the same passage at a slightly faster speed but not yet fluent. Again they would follow the dramatic reading, repeating it several times. Lastly, I would prepare a final tape at a fluent and normal speed. After following the last dramatic rendition several times, all of the students had greatly improved their reading at a fluent level, enunciating each and every word. Many of those students knew those passages by heart. Once we had covered the entire modified play *(Romeo and Juliet)*, which took a few months, the same students would act the play on stage with costumes and background props. During the final presentation, one of the students videotaped their performance. Those children had become fluent with a modified Romeo and Juliet play. I strongly recommend this technique to improve reading, speaking and presenting at the elementary and middle school levels.

Speaking, reading and writing are extremely important during this second myelination *(first and second grades)*. **During this myelin production, all children that have mastered the foundation of the mother language** *(English in Canada, U.S.A., Great Britain, Ireland and other countries where English is their mother language)* **can easily learn other languages** *(Spanish, French, Italian, German, Japanese, etc...)*. **Waiting to teach second languages during middle school is a mistake.** During first and second grades, most children who have an excellent mother language foundation can easily learn a second language or a few other languages. During this second myelination, penmanship *(cursive writing and calligraphy)* **is an important exercise for the development of the left hemisphere.** The part of the brain that controls fine motor skills along the Central Fissure *(cursive handwriting, writing numbers, drawing and doodling)* **greatly affects language development in Wernicke's Area and Broca's Area: fine motor exercises not only positively assist speaking, reading and writing but also greatly help the acquisition of mathematics, grammar, history, geography, second languages, science and other subjects learned in first and second grades. Cursive writing or drawing stimulates all these subjects.** Unfortunately, in the past several years most of the 50 states in the U.S. have done away with cursive writing. <u>**This is a major mistake**</u> that will be costly in the development of the left hemisphere of most American children, unless the parents take it upon themselves to practice cursive writing and calligraphy during first and second grades.

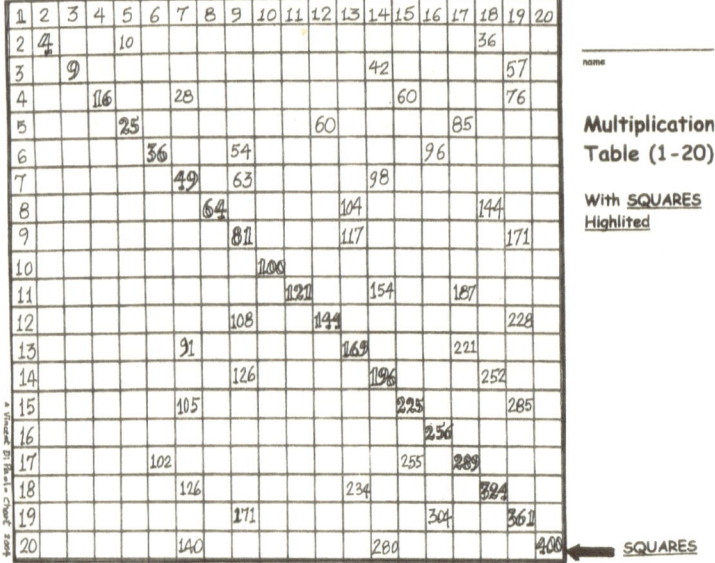

Fig. 17: Multiplication and division table to be used as an exercise page. See Fig. 18 for answers. Educators and parents may photocopy this table and use it with their children. (Vincent L. Di Paolo, 2004.)

During this second myelination in the left hemisphere all elementary school curricula must have a strong foundation of mathematics: multiplication, division, addition and subtraction and their inter-relationship. Unfortunately, there are far too few schools that teach the four operations and their relationship during first and second grades *(the second myelination is also the window to the foundation of mathematics).* **Most schools teach a little addition and subtraction during this time and try to teach multiplication and division from third to fifth grades. Trying to teach new materials like multiplication and division after the myelination is over is very frustrating and extremely difficult because it is too late. All four operations and their relationship must be taught during first and second grades; it is the myelination production that permits this new learning. After trying so many different mathematics programs in the past sixty to seventy years those schools have long given up and started teaching how to use a calculator as early as second grade. <u>Again, this is a very serious mistake</u> because the part of the left hemisphere that should**

be memorizing the multiplication and division table atrophies and the young brain gets hooked on the calculator. Asiatic children are forbidden to use calculators in elementary schools.

In Fig. 17, I have created a multiplication and division table exercise for parents of young children attending first and second grades. As a parent or as a first or second grades teacher you may photocopy this page and use it with your children or students. If your child is in first grade fill the table up to 10. Spend twenty minutes each day working this multiplication and division table with your children of 6 to 8 years of age. If your child is six years old *(first grade)* it is the proper time to begin this year-long exercise. If your child is seven years old *(second grade)* you will have to cover the entire table until 20 x 20. If your child is already in third or fourth grade you may still use this chart but it might take a lot longer and more patience from you as a parent because there is no myelination during these grades. Myelination speeds up learning that is the very reason to practice this chart during ages six to eight while it occurs.

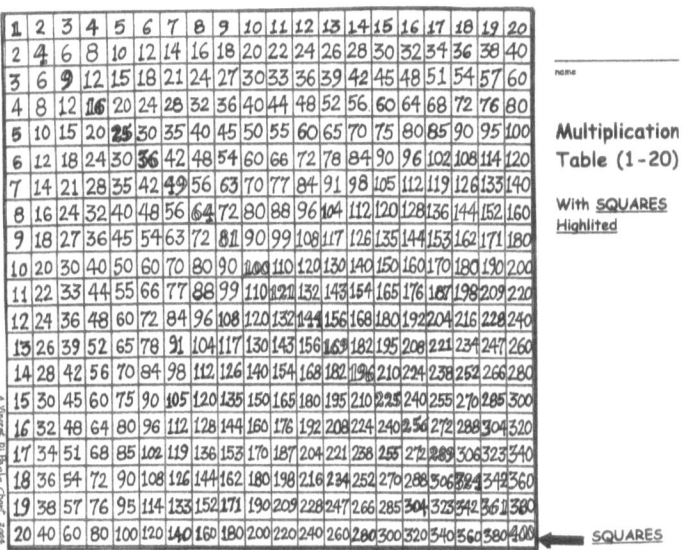

Fig. 18: Complete Multiplication Table for verifying your exercise of Fig. 17. Permission is given to reproduce this chart by parents and educators. Vincent L. Di Paolo, 2004.

Begin by pointing to 1 and say 1 x 1 = 1; 1 x 2 = 2, point to 2; 2 ÷ 1 is 2, point to 2; 2 ÷ 2 = 1, point to 1; 2 x 2 = 4, point to 4; 4 ÷ 2 = 2, point to 2; 4 x 1 = 4, point to 4; 4 ÷ 4 = 1, point to 1; 1 x 3 = 3, point to 3; 2 x 3 = 6, write 6; 3 x 2 = 6, write 6; 6 ÷ 2 = 3, point to 3; 3 x 3 = 9, point to 9; 9 ÷ 3 = 3, point to 3. At this point, introduce the relationship of addition with multiplication and subtraction with division, using manipulatives *(ex.: beans, pasta, colored squares, or any other small objects you may have)* until

3 x 1 = 3; 1 + 1 + 1 = 3; 3 ÷ 3 = 1; 3 − 1 − 1 = 1
3 x 2 = 6; 3 + 3 = 6; 2 + 2 + 2 = 6; 6 ÷ 3 = 2; 6 ÷ 2 = 3; 6 − 2 − 2 = 2
3 x 3 = 9; 3 + 3 + 3 = 9; 9 ÷ 3 = 3; 9 − 3 − 3 = 3
4 x 4 = 16; 4 + 4 + 4 + 4 = 16; 16 ÷ 4 = 4; 16 − 4 − 4 − 4 = 4

you have reached 10 x 10 = 100; 10 + 10 + 10 + 10 + 10 +10 + 10 + 10 + 10 + 10 = 100; 100 ÷ 10 = 10; 100 −10 −10 −10 −10 −10 −10 −10 −10 −10 = 10. Show your child, through this chart, that adding is a slow form of multiplication and subtraction is a slow form of division *(10 + 10 = 20, 10 x 2 = 20, 20 ÷ 2 = 10, 20 ÷ 10 = 2, 20 ÷ 4 = 5 etc...).* Review up to 10 x 10 with your child during the entire first grade by playing little games with this chart. This constant review should be done orally and in writing: writing each and every number *(in both numerical and written forms)* will solidify the memory of it. If your first grade child learns this first half of the table in a few months continue the table until 20 x 20. Practice during the summer following first grade and during second grade until the entire chart is totally memorized. Myelin doesn't stop during long vacations; it is so vital to your child's neurological development that you continue during the vacations after first and second grades. Once your second grader knows the entire multiplication and division chart and knows that multiplication is a fast addition and vice versa and that division is a fast subtraction and vice versa he/she will find mathematics, geometry, algebra and all the sciences much easier to acquire.

In Fig. 19, a chart of perfect squares will help young mathematicians to excel and advance in mathematics and the sciences without the help of a calculator. Looking at this chart will definitely show the magic of mathematics.

As a child, mathematics and geometry fascinated me as I sketched buildings, using geometric shapes and mathematical calculations. As early as five years of age, I was sketching the façades of churches and ancient archaeological structures in my native Cansano, on the Apennine Mountains of Central Italy.

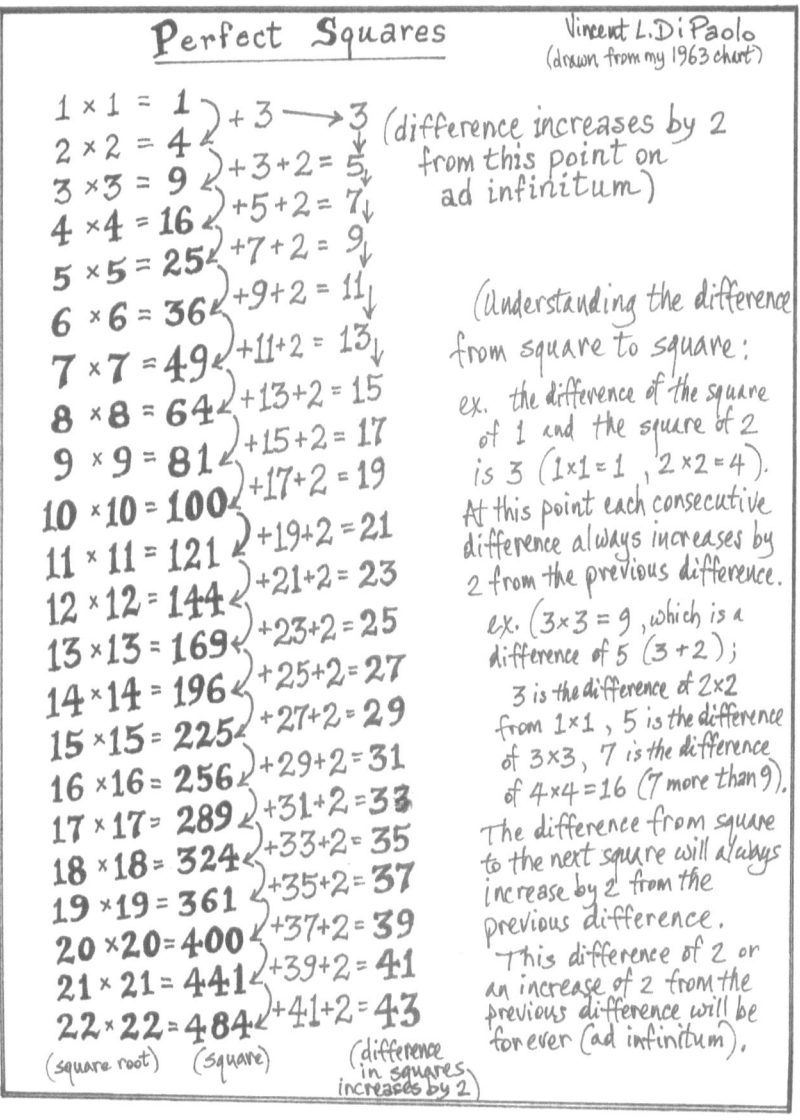

Fig. 19: Chart of perfect squares, which follow a definite pattern. Permission is given to be reproduced by educators and parents. Vincent L. Di Paolo, 2005.

Reading, cursive writing, drawing *(includes sketching and doodling)*, and learning the multiplication table will greatly stimulate Wernicke's and Broca's areas in the Left Hemisphere. As an educator, scientist and artist for the past 49 years I cannot emphasize enough the importance of reading, cursive writing, drawing and learning the multiplication table by heart during this second and final myelination of the Left Hemisphere *(6 to 8.5 years old)*: it will guarantee a success in all subjects, including all advanced mathematics and sciences.

After the second myelination there is a second plateau *(from 8 – 10.5 years old in girls and from 8.5 – 11 years old in boys)* **which begins in third grade, through fourth grade and the first half of fifth grade. Trying to teach multiplication and division during this plateau is very difficult and tedious because that window of learning ended in the autumn following second grade. Also trying to teach mathematical word problems during this plateau is extremely difficult because they require synthesizing multiple information, a right hemispherical function, which has had no myelination during ninety percent of the elementary grades. During third, fourth and the first half of fifth grades plane geometry plays an extremely important role in the developing left hemisphere:** *(if the child has learned it)* **in learning the foundations of geometry and pre-algebra. Plane geometry is visual and its mathematics is quite simple; also, drawing the geometric forms is fun and becomes a powerful foundation for advanced mathematics and sciences.**

Third and fourth graders will be amazed by the pattern of squares and of cubes *(see Figs. 19 and 20)*. **During this second half of the elementary years reading, writing, grammar, history, geography and earth sciences become easier if your child knows the multiplication table. All subjects are inter-related; and, mathematics, drawing, and all of the fine arts glue them together, making it more fun and easier to learn. This type of curriculum follows Jean Piaget's vision of education, which is aligned and synchronized with today's brain and neurological development.**

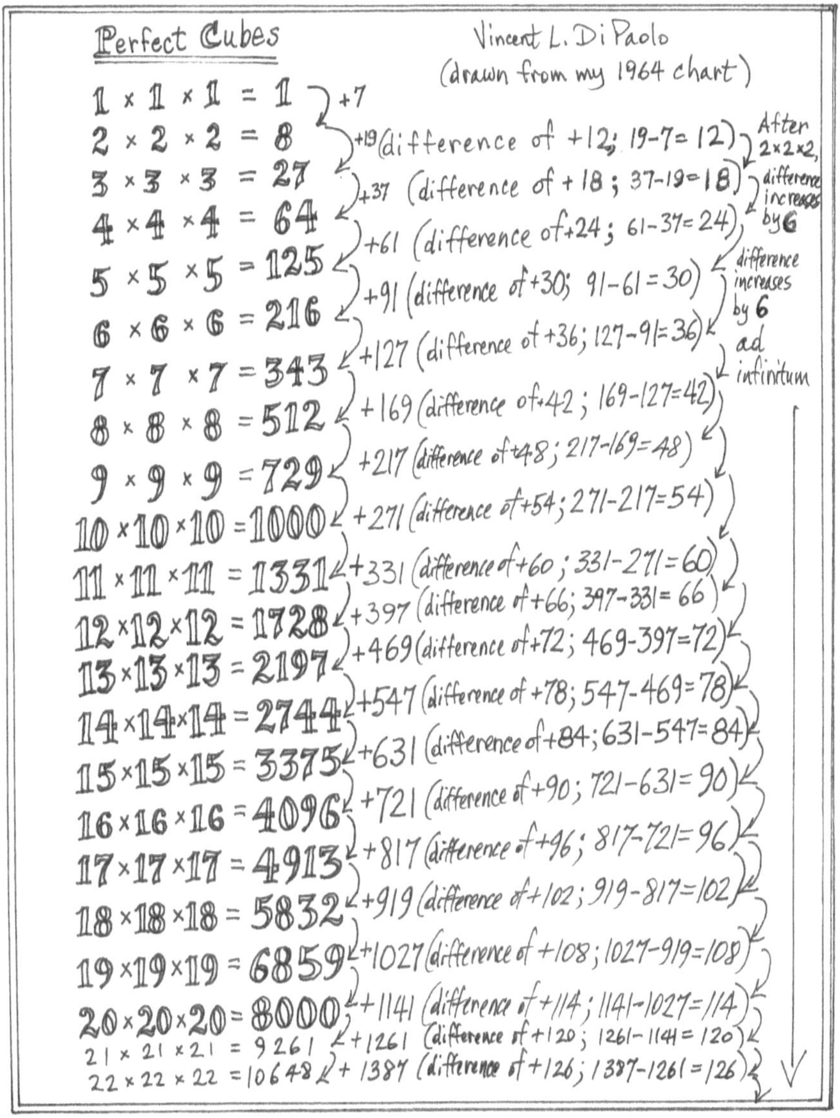

Perfect Cubes

Vincent L. Di Paolo
(drawn from my 1964 chart)

$1 \times 1 \times 1 = 1$ $\quad +7$
$2 \times 2 \times 2 = 8$ $\quad +19$ (difference of +12; 19−7=12)
$3 \times 3 \times 3 = 27$ $\quad +37$ (difference of +18; 37−19=18)
$4 \times 4 \times 4 = 64$ $\quad +61$ (difference of +24; 61−37=24)
$5 \times 5 \times 5 = 125$ $\quad +91$ (difference of +30; 91−61=30)
$6 \times 6 \times 6 = 216$ $\quad +127$ (difference of +36; 127−91=36)
$7 \times 7 \times 7 = 343$ $\quad +169$ (difference of +42; 169−127=42)
$8 \times 8 \times 8 = 512$ $\quad +217$ (difference of +48; 217−169=48)
$9 \times 9 \times 9 = 729$ $\quad +271$ (difference of +54; 271−217=54)
$10 \times 10 \times 10 = 1000$ $\quad +331$ (difference of +60; 331−271=60)
$11 \times 11 \times 11 = 1331$ $\quad +397$ (difference of +66; 397−331=66)
$12 \times 12 \times 12 = 1728$ $\quad +469$ (difference of +72; 469−397=72)
$13 \times 13 \times 13 = 2197$ $\quad +547$ (difference of +78; 547−469=78)
$14 \times 14 \times 14 = 2744$ $\quad +631$ (difference of +84; 631−547=84)
$15 \times 15 \times 15 = 3375$ $\quad +721$ (difference of +90; 721−631=90)
$16 \times 16 \times 16 = 4096$ $\quad +817$ (difference of +96; 817−721=96)
$17 \times 17 \times 17 = 4913$ $\quad +919$ (difference of +102; 919−817=102)
$18 \times 18 \times 18 = 5832$ $\quad +1027$ (difference of +108; 1027−919=108)
$19 \times 19 \times 19 = 6859$ $\quad +1141$ (difference of +114; 1141−1027=114)
$20 \times 20 \times 20 = 8000$ $\quad +1261$ (difference of +120; 1261−1141=120)
$21 \times 21 \times 21 = 9261$ $\quad +1387$ (difference of +126; 1387−1261=126)
$22 \times 22 \times 22 = 10648$

After $2 \times 2 \times 2$, difference increases by 6

difference increases by 6

ad infinitum

Fig. 20: A chart of perfect cubes. Permission is given to be reproduced by educators and parents. Vincent L. Di Paolo, 2005.

CHAPTER 11
MYELINATIONS III & IV: INSULATION OF THE RIGHT HEMISPHERE

By the beginning of fifth grade, the human brain is imbalanced: approximately an average of 110 grams *(105-114 grams = a quarter of a pound)* of myelin have coated the left hemisphere during first and second grades *(during the 2nd myelination)* leaving the brain imbalanced and unstable in the right hemisphere. The *third myelination* begins in the right hemisphere during the second half of fifth grade, continuing throughout most of seventh grade. However, the production of myelin greatly differs between boys and girls directly related to their vastly differentiated hormonal productions.

During the following 28-30 months, *the third myelination occurs* *(10.5 – 13.5 years old)*. A mild myelin production *(approximately 50-70 grams)* in the right hemispheres of boys, directly related to a lesser production of male hormones, mildly insulates their right hemispheres. Finally boys begin to synthesize information *(a right hemispheric function)*, making it easier to solve mathematical word problems, better creative writing and the beginning of emotional feelings and little caring for others *(minimal compared to the girls' emotional feelings and altruistic care)*. However, the production of myelin in the girls' right hemisphere is 300% *(150-200 grams)* compared to the boys' production. This huge difference in the myelin production in the girls' right hemispheres is directly affected, coordinated, and related with their major production of female growth hormones. By 13.5 years old, girls have physically developed into young ladies and their right hemispheres are nearly insulated as their left hemispheres, resulting in almost equally balanced hemispheres and a young adult brain. Meanwhile, the smaller myelination of boys' right hemispheres and their smaller hormonal production leaves them with an imbalanced brain, which directly promotes their childish sophomoric behavior coupled by an uncaring way towards others. By the end of this major myelination, girls have become emotionally young adults and have very strong altruistic feelings towards others.

Most educational systems in the Americas and in most parts of the western world have the same curriculum for boys and girls during secondary grades *(sixth through twelfth grades)* due to the misguided belief that boys and girls develop neurologically the same. However, as some of you *(educators and parents)* by now know that their neurological developments *(because of differentiated myelinations)* are very different. The present curriculum in North America is based on boys' neurological development, cheating the girls out of a possible better synchronized curriculum. During sixth, seventh and eighth grades, girls can learn more advanced materials *(because of their huge myelination in the right hemisphere)* than the boys. During these three grades gender curricula will synchronize learning to the differentiated myelinations between boys and girls: girls can easily learn more advanced mathematics, sciences, social studies and language arts, especially in creative writing *(girls' ability to synthesize information is 300% that of the boys)*, while the boys can follow the present curriculum. Boys and girls may be together in the arts and physical education classes during these middle school grades.

During ninth and the first half of tenth grade there is no myelin production in both boys and girls. This period is another *plateau (periods between myelinations)*. During this plateau it so important not to introduce more difficult subjects, especially during ninth grade, a transition period that results in the most number of failures throughout North American schools. Ninth grade should be a review year of all the information that both boys and girls acquired during the third myelination *(sixth through eighth grades)*. This is a period that should be used to solidify all the materials acquired during middle schools. Presently, more difficult and advanced subjects are introduced in ninth grade and a meaningful percentage of students, especially boys, fail to learn them during a difficult *plateau* in their right hemispheres.

The *fourth myelination* begins in the second half of tenth grade and goes through the end of twelfth grade. During these next two and a half years the girls' right hemispheres receive the last but mild dosage of myelin *(50-70 grams)* while the boys receive 300% of the girls' myelin *(150-200 grams)*. During this myelination *(15.5 – 18.5 years old)* the boys get a huge production of hormones which directly affects their heavy myelination, while the girls receive their last coating of myelin. The boys finally excel in advanced mathematics, sciences, language arts, social studies, creative writing and fine arts. Also, one of their biggest gains is in physical education because they are now young men with the bodies of young athletes. By the end of this myelination both boys and girls have an equally developed neurological system and brain. However, with the present curriculum, the greatest loss in learning is with the girls because they have been following the boys' curriculum. It is so important for school boards to synchronize differentiated curricula with the differentiated neurological growths of boys and girls. If the American educational system would make these important changes our scores in mathematics, sciences and writing would be on the very top.

Another major change needed in North American education that is vital to its future is a separate vocational curriculum that would guarantee jobs to young people specializing in electricity, plumbing, mechanics, carpentry and many other similar fields. Approximately one third of North American students does not succeed with the present curriculum and does not do well in state tests *(like SOL tests)*. Most of these students have the type of brain that specializes in building and mechanics. They would greatly succeed in a vocational curriculum that would give them a high school diploma in electricity, in mechanics, in plumbing and in many other fields. We badly need young people in vocational fields that would guarantee them a very good salary and the possibility of eventually starting their own private companies. Not everyone needs to go to university, especially as their fees have drastically gone up creating a financial burden to parents and students. Also, very much like Europe, professional positions are becoming more difficult to get, except in teaching where more educators are needed.

CHAPTER 12
NURTURING YOUR BRAIN FOR LIFE

The human brain is very much like a microscopic universe, with more than a hundred billion neurons interconnected by possibly trillions of dendrites and synapses formed during early childhood and continued through the myelination periods. Each brain is unique and important; and, we are so lucky to have this most wonderful gift from our universe. It is so important for parents-to-be to have a rich diet, free from drugs and alcohol, to ensure the formation of a healthy brain and neurological development during the nine months of pregnancy; and, it is their duty to nourish their babies' brains with excellent nutrition, lots of parental care and love, and daily reading. Excellent nutrition, daily reading, writing and mathematical exercises are a must during the four myelination, and creating an Omega-3 state of being with lots of antioxidants to keep the brain and neurological system healthy and all the passages of the axons of all neurons free for a maximum and speedy neurotransmitters' activities. Too much omega-6 *(from animal fat)* blocks many of the passages of the axons of neurons, stopping or slowing down the neurotransmitters' work *(bringing new information to memory libraries, retrieving information for writing or testing, and for all daily functions throughout the brain and all its neurological extremities)*. A heavy Omega-6 state in children will cause obesity and a slowing down of learning because neurotransmitters are being blocked or slowed down by the amount of omega-6 in the walls of the interior of the axons of the neurons. A healthy nutrition, rich with vitamin Bs, Omega-3 oil *(olive oil, flax seeds oil, etc...)*, and antioxidants will lubricate the interior of the axons and of the entire vast network of blood vessels, guaranteeing speedy functions of the blood system and of neurotransmitters working towards the acquisition of information during the four myelinations *(from 18 months up to 18-19 years old)* and for the rest of our lives.

We have reached a higher level of thinking which permits us to analyze which food are good for us, allowing us to stay young for much longer; and, we have come to the realization of which foods are harmful to our health and age us quicker. From age twenty until the end of our lives it is important to keep a healthy diet *(Omega-3 and antioxidants state of being)*, physical daily exercises, daily reading and writing, mathematical exercises and drawing *(including doodling)* to guarantee an excellent maintenance of our brain and neurological system.

A heavy Omega-6 state of being and a deficiency of the vitamin Bs *(B1 to B12)* will cause malfunctioning of our organs and the blockage of the axons of our neurons resulting in cancer, tumors, memory loss and eventually dementia, shortening our very lives. I cannot emphasize enough the importance of daily reading and writing, which keep our memory fresh and allow the neurotransmitters to retrieve it at incredible speeds. As you get older, read daily *(newspapers, magazines, novels, scientific books, crossword puzzles or anything that interests you)*, write daily notes, keep a ledger of your daily spending and daily accounting; and, pay all your bills by checks. These daily mental activities guarantee a mental and mathematical exercise which prolongs a healthy memory and a longer healthier life.

CHAPTER 13
INHERITED COSMIC INTELLIGENCE

The human brain is the most advanced development of planet Earth and one of the many cosmic brains in the Milky Way and our universe. We belong to this wonderful planet, Earth, which is one of many planets with intelligence within our galaxy. We most probably share a high percentage of our DNA and intelligence with intelligent beings of other planets within the Milky Way. Also, we most probably share a significant amount of our DNA and intelligence with all other intelligent beings in all the fertile planets of all the galaxies within our universe and within the entire cosmos. Our DNA and intelligence within this solar system that began approximately five billion years ago, when it was formed in the suburbs of the Milky Way, came from hundreds of millions and possibly billions of old star systems that were recycled at the end of their lives by the formidable galactic black hole that rules our galaxy.

Five billion years ago, after the recycling of millions of those older star systems, our galactic black hole spewed out an astronomical number of subatomic particles, each holding one bit of information of all those recycled stars into the dark cosmos. The powerful gravitational force of our galactic black hole pulled back each and every particle as they joined existing nebulae or formed new ones around our Milky Way. After millions of years of cooling down in the colder suburbs of our galaxy a cosmic number of mixed subatomic particles from millions of recycled stars, holding mixed information, began to form our sun and eventually our solar system, along with millions of other new stars. Planet Earth was one of the lucky ones in our solar system to be in the right distance from the sun to become fertile *(containing all the right elements)* and eventually develop life and intelligence from that mixture of DNA and information of millions of recycled star systems, which held previous intelligences seven to ten billion years ago. All the inventions, from the first stone tools to today's technology, we inherited it all from our ancestors that lived in other planets of millions of star systems several billion years ago; and, all the future advanced technology that we will develop in the next millennia that are now dormant in our brains and DNA will be realized as we continue our quest to explore, create and expand our intelligence.

Five billion years from now our solar system *(with our Earth and all the other planets and moons)* and millions of other star systems will have reached the center bulge of the Milky Way, ready to be accepted into our galactic black hole and pulverized into subatomic particles *(each particle holding one bit of information of our planetary system, which includes every living being that ever existed on earth, and all other intelligent planetary systems of other stars).* When those astronomical number of subatomic particles will be spewed into the nearby cosmos and finally form a new nebula or join other nebulae around our galaxy; they will mix and finally begin to reassemble when they reach the right temperature and begin to form new star systems. The lucky planets *(at the right distance from their stars containing all the right elements)* will begin to develop life and intelligent beings. Those future intelligent beings will have inherited our DNA, our intelligence and information, together with intelligence and information from other living planets of other star systems. Eight to ten billion years from now there will be a planet that might look very much like earth with intelligent beings very much like us; and, one of them will most probably be writing a similar book with the same vision as I have and that I have decided to share with all of you, my cosmic brothers and sisters.

CHAPTER 14

EXPANDING COSMOS, EXPANDING COSMIC BRAIN

The human brain is a direct result of the creation of the cosmos and is a microscopic reflection of it, programmed by the Supreme Being to expand in its microcosmic state as the cosmos expands in its limitless malleable macrocosm. As long as we are safe and keep a healthy diet, living in this ever changing gorgeous planet, our brain will continue to evolve as its intelligence will continue to awaken from our dormant capacious indefinable ever-expanding DNA. Our scientific right brain will continue to find ways to prolong our human lives as long as we do not annihilate ourselves. Presently, an average intelligent person's brain *(98-105 I.Q.)* will have employed approximately six quintillion bits of information during 80 years of life, using ten to twelve percent of his/her brain capacity. An extremely intelligent person *(160-220 I.Q.)* will have easily doubled the number of bits of information to twelve quintillions or more *(12,000,000,000,000,000,000)* by only using twenty percent of our present brain capability, leaving eighty percent for future neurological development and expansion. Of course, these numbers are based on today's knowledge of our human brain. The probability that as our brain evolves and expands its capability will also expand is a wonderful phenomenon: that would mean that its capacity of intelligence will keep expanding throughout mankind's future.

I envision, in the distance future, a great development and expansion of our cosmic brain, whose intelligence will expand to a formidable neurological culmination that will allow us to reach the farthest intelligent planets of our galaxy; and, we will be able to develop realistic means to possibly reach other galaxies and explore other intelligent planets, some very similar to ours and some that might be quite different and unique in their developments. I have no doubt that this will happen within the next thousand years. My only regret is that I will not be here to witness those great scientific progressions: even though my brain remains extremely sharp, lucid and young I feel my body paying heavy tolls from all the sports and dare-devil feats that I decided to exercise during my youth, coupled from the ever-consuming power of earth's gravity.

As mankind's intelligence and brain capability will continue to expand our scientific discoveries and inventions will be nearly miraculous; yet, our beautiful planet will go through formidable changes that will have profound consequences in our daily lives. As our intelligence increases in the future we will get a little closer to the divine intelligence of the Supreme Being. I firmly believe that was the Supreme Being's original purpose in programming the development of intelligence throughout the ever-expanding cosmos. When that intelligence will reach total capability in a trillion years or more through the constant recycling of that divine information the cosmos will come to an end as that total divine intelligence will have reached the initial Supreme Being's state; and, possibly another First Big Bang will reoccur.

EPILOGUE

I have tried to put into words for you with the help of a few illustrations the visions that have occupied my right hemisphere of my brain since I was a toddler. It has not been easy to find the proper words to illustrate those visions; but, I believe that my illustrations have facilitated a clearer picture for most of you. As most educators do, I have repeated some of the information in various contextual forms only to help you understand my visions or to strengthen a previous part of this book. Also, I have decided to rewrite my original writing, which is much longer and more difficult to understand, into three smaller books. *"Cosmic Visions Within the Microcosm of My Right Hemisphere: a new theory on the functions of black holes and the development of the cosmic brain"* is the first of this trilogy. You might ask "Why has it taken me this long to write it?" There are a few reasons why I waited sixty-nine years before finalizing writing my visions as a book. Some of the information found in this book I wrote as notes as early as 1963. The charts of Perfect Squares and Perfect Cubes I wrote them in 1963-64 in high school; but I rewrote them more legibly for this book. As a student *(from 1955 to 1970)*, I spent so much time finding ways to improve my understanding of mathematics and geometry by making charts, geometric drawings and paintings; and, I wrote some of my visions on hundreds of single pages.

After graduating with my first two degrees, I began teaching at a high school at the young age of twenty. The following year I married a beautiful young lady, who became the mother of our two children. I was too busy working at two jobs simultaneously: teaching at schools and running a martial arts school. Also, I painted hundreds of paintings and I had a total of 72 art shows during my life. So, as you can see, I did not have any time to write the visions that constantly inhabited my mind. It became more difficult to find time to write this theory once my two children were born as I spent any free time reading to them, playing games with them and creating real adventures for them as I searched for Meso American Indian ruins in Central America, South America and the Southwest of the United States. We did spent some time at beaches; but, I did prefer visiting archaeological sites or trying to discover some in the Central American jungles. I have dragged my ex-wife and my two children deep in the Central American jungles miles away from our vehicles to reach very remote Mayan ruins. We reached one of those ruins by driving nearly twenty kilometers on a dry rocky river bed; and, another time we drove on a donkey path for over sixty kilometers, which took us over twelve hour, finally reaching a small village at midnight. That rental car was not drivable anymore. Luckily, we are all still alive!

I continued teaching physical sciences, algebra, geometry, World History and Geography and fine arts until I was sixty-five. Also, for a long time I was concerned about writing this book because I felt that it interfered with the foundations of my Christian beliefs; however, I now feel that they parallel Christ's teachings in a scientific way. My ex-wife and children gone, I finally began to rewrite about four hundred and forty-six pages of notes, drawings and formulas into three smaller books. I did remove most of the mathematical formulas and sketched and painted illustrations to make it easier to understand my visions, which are very vivid and only visual in my brain.

As an artist and writer it is easier for me to believe in a Supreme Being. However, as you most probably know that most scientists do not believe that there is a Supreme Being; and, many of them believe that the universe developed from nothing. I find that to be so asinine and offensive to my intelligence that I don't understand how so many bright scientists do not believe or see the stunning and evident presence of the Supreme Being everywhere. I do firmly believe in a Supreme Being because I can see its presence each moment of each day of my life. The Supreme Being is the total DNA, all the cosmic intelligence and pure energy found in us, in every intelligent being, in each and every animal, in all plants, in any cosmic body or in any grain of sand, and in every subatomic particle of the entire cosmos. Every cosmic body or subatomic particle has been programmed to be part of the Supreme Matter. Nothing is ever lost; each and every subatomic particle is recycled into a part of a new existence. None of the information or intelligence is destroyed; it is reused into the make-up of a new life *(includes plants, animals, people, planets, stars and each and every cosmic entity)*.

As a Christian *(a follower of Jesus of Nazareth's philosophy of life)* I try my best to be a loving, giving, helpful person; and, I do believe in Pierre Teilhard de Chardin's **anonymous Christian** *(de Chardin firmly believed that any person of other religions who unknowingly follows the philosophy of life taught by Jesus of Nazareth is an anonymous Christian)*. Most people living on earth are good people; but, there are groups of people from every walk of life that do not believe in peace and love for mankind. Today, we have so many wars, so much destruction and so much famine because of those bellicose people, who do not believe in giving, sharing and love; rather they have made money their god and have spent their lives in developing many extremely lucrative businesses that manufacture and sell weapons and other war machineries. After all, Jesus did say that we are all children of God; and, we are all part of the Supreme Matter. Christ said that there is life after death: every subatomic particle that shapes us as human beings will be reused into the make-up of other intelligent beings in the distant cosmic future.

It might be very difficult for many of you to accept my theory of a Supreme Matter, of the functions of black holes, and of the universal development of the cosmic brain in trillions and very possible in quadrillions or quintillions of earth-like fertile planets throughout our known universe; but, I am merely illustrating my long time visions that stay constantly vivid in my mind. Also, as we progress in science and technology, it would be more beneficial for us, for our children and grandchildren, as humans, to clean up and rejuvenate our beautiful planet rather than try to colonize the unforgivable and unlivable planet Mars. In a billion years from now, as the sun enlarges, planet earth will become unlivable *(probably very much like Venus is today)* and no one will be able to survive here; but, Mars most probably will become very much like a primordial earth. Only then it might be habitable!

I want to finish this book with the following message: I thank the Supreme Being every morning for giving me another day in this extraordinary beautiful wondrous planet Earth as I hear so many gorgeous birds chirping in my backyard as they feed on their daily supply I place in their bird houses. I love all animals and plants because it is our duty, as intelligent beings, to be the guardians of all living creatures and natural resources that inhabit this wondrous planet.

We are so lucky to be alive for our lives are so precious and important. Each one of you should be thankful for each new day of your lives. You must believe in yourself, in your dreams, and in the goals that you chose to set for your future. You have to work hard each and every day to have those dreams become reality, never giving up especially when things go wrong, even when you lose the people that you love or your precious possessions. Always be courageous and start working toward your dreams again. Never judge anyone and accept each and every person for who they are, accepting their individuality and the life they have chosen no matter how different or foreign they are from yours. Each person is unique and possesses the ability to reach his/her dreams. Be kind and generous; be willing to share with others, even total strangers. Do not bully anyone because you must know how bad it feels to be bullied. Do something good for others each and every day: it doesn't have to be something big. Greet your neighbors and strangers daily! If you have a family and have children tell them "I love you!" each and every day. Hug and kiss your children so they feel loved; and, hug and love the person you have chosen to be your partner and friend for life. Read a children's book to your child or children each and every night: spend that half hour with them before they go to sleep and whisper "I love you!" in their ears as you kiss them good night. Believe me; it will make a big and positive difference in their young lives and in the development of their precious brains.

We the people of this Earth are supposed to be the keepers and guardians of all living creatures: animals, plants and every geological

object *(even rocks are living and have cosmic energy)*. **Do not be mean to animals and plants and do not destroy them; be gentle with them for they will feel your kindness.** If you see a paper, or plastic bag or bottle or anything that has been thrown by someone else pick them up and recycle them. Imagine if all of seven and three quarter billion people that inhabit this planet would pick-up a few objects they find each and every day how quickly we would clean up our beautiful Earth. We take so much from this planet but we are so careless in how we keep it. Make a pledge that from now on you will be nice, generous and understanding with all humans, all animals, all plants and all the gorgeous places of this planet that you visit; and, pledge to pick up anything that has been carelessly thrown away and recycle them. If we can all do this each and every day for the rest of our lives we will clean planet Earth and our lives will surely become very joyful and more meaningful in this little beautiful blue planet in the middle of this ever expanding cosmos.

GLOSSARY

(Words, names and some cosmic and microcosmic numbers related to this book, written in my own words to simplify their meanings and readableness.)

Absolute zero (0°Kelvin = -273.16°C = -459°F) is the temperature at its lowest possible state within the cosmos.

Accretion of a Galactic Black Hole *(during formation)* is the accumulation of stellar black holes *(formed at the end of their lives)* that have reached the center of the young galaxy *(after 9.5-10 billion years)* pulled by the powerful galactic gravity and resulting in the formation of a young galactic black hole *(see Fig.11)*.

Acetylcholine is a neurotransmitter (organic chemical $C_7NH_{16}O_2+$) that helps the acquisition of information (learning), memory, REM and several other functions including regulating the endocrine system and muscular and skeletal contractions. A deficiency of acetylcholine can lead to muscular and skeletal problems.

Alexion is a new word that I have taken from the Greek word ***alexi***, which means to protect. In my new theory ***alexion*** is a protective coating given to each subatomic particle from the very beginning of time by the Supreme Being to protect them from any temperature and from the most powerful winds generated in black holes. Alexion will always protect each subatomic particle forever or until the end of the cosmos.

Andromeda Galaxy is the nearest major galaxy from the Milky Way; it is a huge spiral galaxy holding a possible one trillion star systems. In three to four billion years from now Andromeda will collide with the Milky Way to form a super mega galaxy.

Angstrom (Å) is an astronomical unit, which equals to 10^{-8}cm, named after Anders Ångström, a Swedish astronomer and physicist born in 1814 and died in 1874.

Antioxidants are super chemicals, found in fruits and vegetables that attack and destroy oxygen-free radical agents that live inside our bodies, including all mammals. These dangerous radical agents can develop into cancerous tumors by constantly feeding on Omega-6 *(animal fat in meat)*; however, a rich diet of antioxidants will break down and eventually destroy them. Some of the most powerful foods that contain antioxidants are: *(in order of potency)* prunes, raisins, blueberries, blackberries, garlic, kale, strawberries, and all fruits and vegetables. Antioxidants also help our neurological system by zapping radical agents traveling through it.

Apennines (Central) is the central section of the Apennine chain of mountains that begin from above Genoa and crosses the Italian Booth all the way to the bottom of Calabria. It is a volcanic chain of mountains with several volcanic eruptions and earthquakes throughout its history.

Aquila is a beautiful medieval city and the capital of the province of Abruzzi in Central Italy. It is two hours east of Rome by car and one hour from beautiful beaches of the Adriatic Sea.

Asteroids are found in the Asteroid Belt, between Mars and Jupiter. There are nearly 600,000 asteroids within the belt: the biggest is Ceres *(950 kms. diameter)*, which looks like a moon. The rest are shaped like rocks, some as large as 400-580 kilometers in diameter and some as small as a boulder. Some asteroids have escaped the belt and threaten to hit Mars and possibly Earth.

Atom is composed of a nucleus, containing protons and neutrons, circled by electrons. Hydrogen is the simplest atom with only a proton (+) in its nucleus and an electron (-) circling it. However the simple hydrogen atom is the major make up of the entire cosmos. All the other atomic elements are created in healthy stars through the fusion of hydrogen.

Atomic mass unit (amu) is the SI unit that expresses the masses of subatomic particles in all atoms. Both protons and neutrons have a mass of 1 amu. However, electrons are so small that it would take approximately 1,810 electrons to equal the mass of 1 proton or neutron. Therefore, the mass of an electron is so insignificant that it equals zero.

AU is the astronomical unit which equals to 1.495979×10^{11}m. = 149,597,870,700 ms. = approximately 150,000,000 kilometers = approximately 93,000,000 miles, average distance from Earth to the sun.

Average speed is the total distance divided by the total time (ex. Average speed of earth spinning on its axes = 40,010 km. *(circumference of earth)* ÷ 24 hrs. = 1,666.7 km/hr).

Black Holes are the super recycling machines of the cosmos that break down star systems into subatomic particles and shoot that astronomical number of stellar particles into the cosmos to later regroup into nebulae and eventually form new stars. I cannot agree with the theory that stellar black holes evaporate and disappear, leaving no trace. However, in this book, which states a new theory, stellar black holes disappear, leaving new nebulae containing all the subatomic particles that had previously composed each one of those stars. There are **stellar black holes, galactic black holes and cosmic black holes.** The functions of black holes guarantee a continuity of creation by the recycling of old star systems into nebulae which eventually will regroup, mixed with other existing subatomic particles, into new stars. The speed of the winds generated inside a galactic black hole can reach 250,000 km/s., or 907,200,000 km/hr., which is about 84% the speed of light. Many astrophysicists and cosmologists believe that the speed of cosmic black holes may reach a maximum of 99.531% the speed of light or 298,387 km/s. Nothing is wasted in the cosmos through the work of black holes. During pulverization of star systems the temperature inside a galactic black hole can reach a billion degrees Celsius and possibly more.

Bocono River is a tributary in the province of Trujillo, Venezuela, that empties into the Portuguesa River, which empties into the Apure, which empties into the Orinoco River.

Broca's Area is a small but very important section in the lower part of the frontal lobe of the left hemisphere of the human brain. It specializes in language, especially speech, oral presentations and enunciation.

Brown dwarf are very small stars that cannot fuse enough hydrogen into helium stunting their stellar growth. I firmly believe that Jupiter is a brown dwarf because it has all the characteristics of a brown dwarf and was able to fuse only about 10.5 % helium. It looks like a baby star with a large number of satellites, some of which contain water. Brown dwarfs are between the mass of Jupiter and 0.0999M *(M=solar mass)*.

Byzantine Empire was named after the ancient city of Byzantium, which was renamed Constantinople when Emperor Constantine made it the Eastern Roman capitol in 330 A.D. The Roman Empire had been already split into two by Emperor Diocletian in 293 A.D.: the Western and Eastern Roman Empire. It was Constantine that named it the Byzantine Empire, which lasted until 1453 A.D. when it fell to the Turks and Constantinople was renamed Istanbul.

Cambrian evolutionary explosion was the development of a multitude of animals during the Cambrian Period on planet Earth (600 million years ago during the Paleozoic Era).

Cansano is a small medieval town, built over the ancient pre-Roman Ocriticum, in the province of Aquila, Abruzzi, in the Central Apennines region.

Central bulge is also called the Galactic bulge in the center or nucleus of the Milky Way and most similar galaxies. It is made up of hundreds of millions and possibly billions of older stars that have reached the end of their lives and circle around the galactic black hole, ready to be recycled periodically.

Cerebellum means *little brain* in Latin. It is found at the back of the human brain, under the occipital lobes. It has two hemispheres, made up of fine folds *(folia)* and subdivided into three lobes. The cerebellum subconsciously controls all our movements, muscles, balance and posture.

Chandrasekhar mass *(from Subrahmanyan Chandrasekhar's Theory)* is a limited mass of any star, equal to our sun or up to 1.4399M = *(2.765 x 10^{30}kg.)*, which eventually becomes a white dwarf at the end of their lives

$$\left(M_{limit} = \frac{w_3^0 \sqrt{3\pi}}{2} \frac{hc^{3/2}}{G} \frac{1}{(\mu_e m_H)^2} \right).$$

Colle Mitra is a small mountain outside of Cansano about 5,000 feet in elevation.

Cordillera de Merida is a chain of mountains in the western part of Venezuela, which extends from Valencia *(north)* to San Cristobal *(south)*. Its highest mountain is Pico Bolivar, 5007 meters *(about 16,411 feet)*.

Corpus Callosum is a neurological bridge, made up of approximately 200 million nerve fibers, connecting the left hemisphere to the right hemisphere of the human brain, permitting constant communication between the two.

Cosmos is the entire cosmic space that contains our universe and the possibility of an infinitesimal number of other universes.

Cynodont Therapsids were late reptiles that were developing into mammals, with dog-like teeth and features during the very Late Permian to Early Jurassic periods *(260-175 million years ago)*. Mammals developed from Cynodont Therapsids as their brain acquired a mammalian brain wrapped around their reptilian brain.

Dark Ages *(see Krakatoa)*.

Dark Energy resides within each gravitational center, like the center of the earth, the center of stars, the center of any black hole, and it rules the entire cosmos by controlling the matter of all cosmic bodies and all the dark matter in-between.

Dark matter *(in this book)* is the *cartilage* in between matter, from the smallest space between particles to the huge space between planets, moons, stars, galaxies and universes. The bigger the masses of cosmic objects the more dark matter is needed in between to guarantee protection of those cosmic bodies.

De Chardin, Pierre Teilhard *(1881-1955)* was a French philosopher, paleontologist, geologist and a Jesuit priest. In one of his books, he coined the term *anonymous Christian*.

Degenerate stars are all the neutron stars and white dwarfs, whose interiors are under extremely very high pressure, compressing the electrons and atomic nuclei into very tight and heavy matter, resulting into a degenerative state of their normal functions. Chandrasekhar theorized that stars like the sun, with a mass of 1M to a maximum 1.439999 x M *(mass of the sun)* will eventually degenerate into white dwarfs or neutron stars: this limit is known as the *Chandrasekhar mass limit*. Any star with much greater masses will eventually implode into stellar black holes after engulfing and burning all of its planets and moons.

De Saint-Exupéry, Antoine (1900-1944) was a French aviator and writer, famous for his book, "The Little Prince".

Dopamine is a very important neurotransmitter *(organic chemical $C_8H_{11}NO_2$)* that helps to regulate emotion and movement. At an early age, it is dopamine that can get hooked on reading or playing with game gadgets. Lack of dopamine can cause Parkinson's disease, ADD and ADHD while a low dopamine production can lead young people to various addictions. Dopamine is a catecholamine neurotransmitter.

Electrons are the negatively charged tiny particles that circle the nuclei of all atomic elements. In the atomic composition of all elements there are the same numbers of electrons (-) as protons (+); but, the number of neutrons *(no charge)* can sometimes be more than the number of protons *(example: an aluminum atom has 13+protons and 13-electrons circling in its cloud but its atomic mass is 27 because there are 14 neutrons in its nucleus, making aluminum an isotope).*

Epinephrine is a hormone that is manufactured in the adrenal glands *(in the medulla)* to increase cardiac performance during extreme stressful situations. It is also called **adrenaline** and is related to **norepinephrine.**

First Big Bang *(in this book)* is the very first moment that the Supreme Matter exploded at the very beginning of time propelling all of Its intelligence, all of Its information and DNA into the early cosmos, allowing it all to expand, grow and develop into the present and future cosmos.

Fourth Ice Age began approximately 25,000 years ago and it reached its peak at about 18,000 years ago. The warming trend that followed it began about ten thousand years ago and by around 5,600 B.C. *(7,618 years ago)* the oceans had increased by several hundred feet, flooding the fertile pre-Sahara fertile valley and raising the level of the Mediterranean Sea and Black Sea considerably. It took over 2,000 years for the waters to recede, leaving hundreds of feet of sand from the Atlantic Ocean on top of the pre-Sahara valley, forming the Sahara Desert. Presently, we might be towards the end of this warming trend, which eventually will trigger a fifth ice age *(all the melting ice from the Arctic and Antarctic are slowing down ocean currents resulting in the beginning of an ice age in the near future).*

Frontal Lobe is found in the front of the human brain *(both left and right hemispheres)*; it makes the final decisions of all information received and sends them to the cerebellum which controls all actions based on those decisions.

Fusion is the nuclear fusion of two hydrogen atoms into helium in stars; it is also the fusion of helium into heavier elements in healthy stars. A healthy star, through its fusion, will develop all necessary elements to give life in healthy planets like Earth.

Galaxy, (galaxies) are like cosmic platters that holds billions of star systems spinning around its center, which contains a galactic black hole covered by the central bulge, hundred of millions and possibly billions of oldest stars that have reached the end of their lives. The nebulae formed by subatomic particles from the recycling of older stars are the nurseries of new stars in the suburbs of each galaxy, which are formed by a mixed astronomical number of those particles.

Gluon is a subatomic substance that unites quarks to become protons or neutrons. In this book, which is about my vision of the cosmos, ***alexion*** protects quarks and possibly morphs into ***gluon*** at lower temperature above zero in uniting quarks to form protons and neutrons.

Gravity according to Einstein is the consequence of the curvature of spacetime as a result of the uneven distribution of mass *(the cores of cosmic bodies are the heaviest pars of those bodies, made up of mostly iron and nickel, creating a force within those spinning bodies)*. Subatomic particles are bound by microscopic gravity while planets, stars and galaxies are bound by powerful macroscopic gravity. Gravity rules the cosmos by holding all cosmic bodies in their respective spaces, rotations and revolutions. Gravitational potency varies with each different cosmic body, depending on the weight of their cores, which directly affects their rotation velocity *(spin)* and revolution. It is the gargantuan amount of iron in the cores of large stars that allows the development of stellar black holes. Newton's theory of gravity $(F = G \frac{m_1 m_2}{r^2})$ works great on planet earth only, based on earth's gravity.

Homo Neanderthalensis (*neanderthal man*) went through an impressive neocortex development before modern man. However, most Neanderthals were cannibals, a fact based on hundreds of human skulls found on digs of Neanderthal villages which had been carved as bowls used for drinking and eating. The expansion of their skulls might have directly resulted from the daily practice of cannibalism.

Hydrogen atoms (^1H) each contain a positive-charged proton and a negative-charged electron. It is the simplest of all elements but it is used to create all other elements. Hydrogen atoms make up the cosmos; they compose approximately 75% of the baryonic mass of the entire cosmos.

Hydrogen atoms are fused in stars to create helium and all the other elements necessary to create life.

Industrial Revolution began in England during the early 1760s and quickly spread throughout Europe. By the early 1800s it had also spread in the Americas *(developed by European immigrants)* and later in Asia and parts of Africa as some European powers were pursuing to colonize many of their countries. As machines and power tools *(generated by burning carbon)* replaced hand tools in factories large amount of carbons escaped into earth's atmosphere, speeding up air pollution especially in larger industrious cities. I firmly believe that the Industrial Revolution greatly helped to speed up global warming on planet Earth.

Jurassic Period was the second period of the Mesozoic Era *(200-145 million years ago)*; it followed the Triassic Period *(250-200 million years ago)*. It was during the Jurassic Period that Pangaea *(the huge land formation containing all continents together that formed as the planet cooled)* split apart as a result of numerous volcanic explosions that triggered the tectonic plates to move towards today's geographical positions as water poured into the spaces forming our oceans and seas. Continents, mountains and oceans developed during that chaotic period, with developing sea life; simultaneously dinosaurs and early mammals roamed the new land formation. Also, early birds developed from the pterosaurs that roamed the earth's sky.

Krakatoa is a super volcano on the island of Rakata, in the Sundra Strait between Java and Sumatra islands. It erupted in 535 A.D. with an explosion that equaled two billion atomic bombs. Its explosion was heard, felt and recorded around the world; its volcanic ash traveled around the earth creating a dark belt of volcanic ashes which blocked the sun for nearly twenty years, resulting in the Dark Ages, giving rise to the plague and famine. More than half the population of planet Earth died and most of the living were sick and lived in misery. It took several hundred years for the Earth to rejuvenate itself beginning with the Renaissance in Italy, which finally spread throughout Europe.

La Gran Sabana *(the Great Savanna)* begins south of Angel Falls and borders with Brazil and is part of the Guyana highlands and are extremely rich in iron ore, natural gas and diamond deposits. It is also rich in manganese, copper, salt, sulfur, asbestos, coal and bauxite. La Gran Sabana is also rich with wildlife, including mountain lions, jaguars and other big cats.

La Guaira is the port of Caracas, Venezuela, located on the Caribbean Sea.

Left Hemisphere of the human brain is also called the logical brain *(the super neurological secretary)* or the rational brain; it strictly follows all grammatical and mathematical rules. It is the abstract thinker and does all the work necessary in the acquisition of information.

Light year *(ly)* equals to 10^4 AU x 6.324 = 10,000 x 150,000,000 km. x 6.324 = 9,486,000,000,000 km. *(9 trillion 486 billion kilometers or 5 trillion 881 billion 320 million miles).*

Limbic System - *(see Mammalian Brain).*

Mammalian Brain is also known as the Limbic System, wrapped around and neurologically connected to the Reptilian Complex. It controls blood pressure, heart rate, and the production of hormones, myelin and endorphins; it also controls feelings and emotions. It is the part of our brain that nurtures us and acts very much like a mother to the human body.

Mass of our Sun = 1.989109×10^{30}kg $\approx 1.99999 \times 10^{30}$kg.

Matter is anything that has volume and mass, from a grain of sand to mountains, from a flower to animals and people, from the smallest asteroid to the biggest planet or star.

McGill University was chartered in 1821 in Montreal Canada; it is considered a great university and is world renowned in many fields. It is my alma mater and I am proud to have studied there. My daughter, Alexandra has received three degrees from McGill University. There are many American students at McGill, together with students from all over the world.

Middle Ages is a period of the earth's history that lasted about 900 years *(560-1500 A.D.)*; it began after the Dark Ages, through the Carolingian Empire *(Charlemagne)*, feudalism, the Crusades, 100 Years War, ending with the development of the Italian Renaissance as it spread throughout Medieval Europe. The human brain greatly suffered because of the Dark Age and most of the Middle Age due to a cataclysmic explosion of Krakatoa, which poisoned most of the earth, causing the Black Death and other plagues. It took more than 700 years for the Renaissance to happen.

Milky Way *(Lat.: Via Lactea)* is our galaxy, which contains more than 300 billion stars; one of those stars is our Sun. There are possibly trillions of planets, many with moons. It also contains beautiful nebulae, where new stars are created from the mixed subatomic particles from billions of recycled older stars. At its center, hidden and surrounded by the central bulge, there is a formidable galactic black hole which constantly recycles older star systems into subatomic particles *(used to create new star systems with new earth-like planets with new intelligent beings)*.

Miocene Period or Epoch *(approximately 23 to 5.3 million years ago)* was the period during which mammals, including whales, and birds ruled the earth. Apes greatly developed during this period and by the last two million years of this period *(approximately 7.5-5.4 million years ago)* hominids separated from other apes. Towards the end of the Miocene Period temperatures dropped as the North Pole began its expansion into an ice age. The Himalayas were formed during this period as India smashed into Southern Asia.

Missing Link is a possible intermediate period in the genetic evolution of humans. The idea of a missing link is not considerate a scientific term because it has not yet been proven. HAR_1 and $FOXP_2$ genes might possibly solve the missing link problem, as they most probably developed new types of RNA *(ribonucleic acid)* and DNA *(deoxyribonucleic acid)* in humans.

Myelin is the fatty insulation produced to wrap around the axons of neurons during myelination periods. I strongly believe that the production of myelin is directly affected by the production of hormones in humans.

Myelin is made up of approximately 80% lipids *(including 15% cholesterol)* and 20 % protein; they are mixed with approximately 40% water. Myelin insulation increases the speed and functions of neurotransmitters as they travel through the vast neurological system *(approximately 421,000 kilometers or 260,000 miles)*. A rich diet will result in a rich production of myelin which permits faster messaging and better learning in children *(1.8-18.5 years old)*. Although there is a small myelin production during the last two months of pregnancy, which is needed for slower messaging, ninety-five percent of myelin will be produced during four distinctive myelination periods *(averaging from 18 months to 18.5 years old)*. It is so import that babies and children continue to drink whole milk to help produce more myelin. Also, a high fatty diet for children helps to produce stronger myelinations and hormonal productions. I firmly believe that low fat milk is not healthy for children for it will result in a lower production of hormones and myelin.

Myelination Periods within the human brain and neurological system begin around 18 months after birth in girls and around 22-24 months in boys. There is a small myelination during pregnancy to allow small and slower messaging for the embryo. At birth, a baby's brain weighs an average of one pound. During the first year of a baby's life the brain doubles in weight *(2 pounds)* as the cerebellum develops to control all muscular and skeletal movements. The final pound of the human brain, which is the productions of myelin, occurs in four stages: the first myelination occurs in the left hemisphere earlier in girls *(18 months-4.5 yrs. old)* and a few months later in boys *(22-24 months-5 yrs. old)*; the second myelination also occurs in the left hemisphere earlier in girls *(6-8 yrs. old)* and about six months later in boys *6.5-8.5 yrs. old)*; the third myelination occurs in the right hemisphere, again a few months earlier and 300% *(of the boys' amount)* in the girls *(10.5-13.5 yrs. old)* and only 100% in the boys *(11-14 yrs. old)*; the fourth myelination occurs in the right hemisphere and is 300% in the boys and 100% in the girls *(15.5-18.5 yrs. old)*. By the end of the fourth myelination both boys and girls have an equally balanced young adult brain and they are ready for higher levels of learning.

Nebula, nebulae are galactic clouds formed by the atomic and subatomic particles spewed from black holes from the recycling of old stars. Nebulae are also known as *giant interstellar molecular clouds* of hydrogen, helium, ionized gases and carbon monoxide. Some galactic nebulae are from 9 to 11 million times the mass of the sun. Nebulae are the nurseries of future stars and future intelligence.

Neocortex is the most developed part of our brain, neurologically interconnected to the mammalian brain and to the deepest parts of the reptilian complex. It is the neocortex that makes us very intelligent mammals, which includes whales and dolphins. The neocortex is divided into four lobes, which make us different from other animals. The neocortex permits humans to be able to speak, read, write, think, theorize and invent.

Norepinephrine is very similar to epinephrine and is manufactured in the adrenal medulla. Exercise raises the amount of norepinephrine and epinephrine, which causes the heart to beat faster increasing blood flow. Norepinephrine helps memory, REM sleep and romantic love. Low levels of it can result in ADHD, aggression, depression, and the lack of it can result in Parkinson's disease.

Nervous System or **Neurological System** controls our body's necessary processes and functions; it also controls and stimulates most of our organs. It includes our brain, spinal cord, ganglia, and all nerves, which controls and co-ordinates all the responses to various stimuli of our senses; it also controls our emotions, and all of our conscious behavior.

Neurotransmitters are chemical messengers produced together with hormones, specializing in messaging throughout the brain and neurological system. Many neurotransmitters specialize in bringing new information to the memory banks or retrieving some of that information, in focusing or paying attention. Vitamins Bs directly affect the production of neurotransmitters and their speed of messaging. The lack of vitamin Bs results in less neurotransmitters and slower speed, attention deficit, ADHD and poor behavior.

Neutrinos were first thought by Wolfgang Pauli in 1930; but it was Enrico Fermi who named them *neutrinos* in the summer of 1932 at a Paris conference. *Neutro (Italian)* means neutral; therefore *neutrino (Italian)* means a tiny neutral subatomic particle. I firmly believe that neutrinos are the remnant tiny subatomic particles discarded in solar or stellar fusion. I also believe that neutrinos are created in stellar black holes, galactic black holes and cosmic black holes as they break down old star systems into subatomic particles, leaving an abundance of tiny subatomic particles *(neutrinos)* floating in galactic nebulae. When the subatomic particles finally begin to reassemble when the nebulae reach the proper low temperature neutrinos play an important part in filling the gaps inside the nuclei of hydrogen atoms and in the nuclei of other elements created by fusion in stars. There are three known types of neutrinos: *Electron neutrino (Ve by Pauli), Muon neutrino (Vµ around 1946-49), and Tau neutrino (VT around 1975-76).* There is a possibility that a fourth much heavier neutrino exists to balance the lightness of these three. If the fourth heavy neutrino exists it would assist the negative force of gravity present in its subatomic state. It is possible that neutrinos might assist gluons in the formation of protons and neutrons.

Neutrons are subatomic particles that have no charge but have approximately the same mass as protons. They are found in all nuclei of all elements except the hydrogen atom, which has only one proton and one electron.

Occipital Lobe is in the hindbrain *(back of our brain)* and is activated by visual stimuli and retains visual information memories. The more complicated and colorful sights strongly stimulate the occipital lobe.

Ocriticum is the ruin of an ancient town *(approximately 2,000 B.C.)* on the outskirt of the medieval and modern town of Cansano, at approximately 2,800 feet elevation in the foothills of the Central Apennines in the province of Aquila, Abruzzi, Italy.

Olympus Mons is the largest and highest shield volcano in Amazonis Planitia of Mars; at 21,287 meters *(69,844 feet)* tall Olympus Mons is the biggest and largest mountain in our entire Solar System and three times taller than Mount Everest.

Omega-3 oils are found in fish, seafood, olive oil, omega-3 gels, and linoleic acids that are found in nuts, flaxseed and green leafy vegetables. Omega-3 oils lubricate the axons of our neurons and the inner walls of our blood vessels, clearing them for better flow and faster messaging. Deficiency in Omega-3 oils can lead to ADD, ADHD, dyslexia, minimum learning and other neurological problems.

Omega-6 is found mostly in animal fat, corn oil, hydrogenated vegetable oils, and in trans fatty acids. Too much Omega-6 fats in your body can eventually accumulate into cancerous tumors *(as free radical agents feed on Omega-6)*, especially around our waist; it also blocks your arteries and the axons of your neurons. If the axons are blocked there is no messaging, no learning, and no retrieval of memory. If you are a meat eater, remove all the skin and fat of any meat you eat; and, take an Omega-3 gel to fight the Omega-6 fats. Also Omega-6 can lead to obesity and diabetes. Control what you eat and make sure that you have more Omega-3 oils in your body and less Omega-6 fats.

Paleocene Period *(approximately 66-56 million years ago)* became much warmer and very humid towards its second half compared to its cooler first half. Most land dinosaurs had been wiped out by the beginning of the Paleocene period, during the Cretaceous Period when a huge asteroid *(approximately 10.5 kilometers wide)* smashed into the Yucatan coast in Mexico, forming the Chicxulub Crater. As the ocean waters became warmer a great variety of marine life developed during this period.

Parc des Hirondelles *(Swallows Park)* is a beautiful park in Montreal with soccer, football, and baseball fields. It has a bocce ball rink and a hockey and skating rinks in winter. There is a hill in the middle of the park with a ski lift, free for everyone to use.

Pantano is a prairie outside of the town of Cansano; it is totally fenced with a gate where you may bring your horses for the day to graze on many nutritious grasses. It is free for citizens of Cansano.

Parietal Lobe lies between the Fissure of Roland and the Sylvian Fissure *(between the frontal and occipital lobes)*; it processes the sense of touch, spatial sense and direction. A few parts of the parietal lobe help with language processing.

Piaget, Jean *(1896-1980)* was a Swiss psychologist who developed *the Theory of Cognitive Development*, which parallels today's modern brain theory and is used by many educators worldwide. He was also the Director of the International Bureau of Education. Most educators throughout the world have studied Piaget and his theory. I attended a Piagetian school in Italy from 1955-1960. Piaget's theory definitely made me a better educator.

Plateau, plateaux is a neurological stop in the productions of myelin, and periods of rest in between myelinations.

Pliocene Epoch *(5.3-2.6 million years ago)* followed the Miocene Epoch. It was during this epoch that *Australopithecus* developed from earlier hominids. By the end of the Pliocene Epoch *Homo habilis* had evolved, using more sophisticated weapons and tools than earlier hominids.

Prince William County Public Schools is the school board for all public schools in Prince William County in Virginia, U.S.A.

Probability of possible life in our universe depends on the masses of stars if M *(mass of star)* is in between 1.5999×10^{30}kg. and 2.3999×10^{30}kg., or if M is in between 1.989109×10^{30}kg. and 1.99999×10^{30}kg. (See my Probability Chart in Chapter 5).

Protogalactic Clouds are vast galactic nebulae created by cosmic black holes, made up of all the subatomic particles from old recycled stars and galaxies. As the protogalactic clouds gain mass, speed and momentum the gravity that develops flattens them into platters or disks, which eventually become spiral galaxies. If there is not enough speed in their spin to form a disk, they will eventually become elliptical galaxies. The gravitational pull from larger galaxies *(ex.: Andromeda)* can eventually cause a collision with smaller or medium galaxies *(ex.: the Milky Way)* to form a mega galaxy while parts of both galaxies are obliterated in the process. The atomic particles from the millions of obliterated stars can possibly form a protogalactic cloud or join the vast network of galactic nebulae.

Protons are positively charged massive particles in the nuclei of all atomic elements. All protons in all atomic elements are identical and each proton has a mass of approximately 1.7×10^{-24} g = 0.00000000000000000000017 g.

Protostars are baby stars developing in all nebulae. Many of them will eventually develop into healthy stars as they continue to gather atomic particles from their maternal nebula. As their small cores accrue more particles their masses increase and their cores become denser and larger resulting in early gravitational development. As the protostars evolve fusion of hydrogen-1 and hydrogen deuterium begins, creating helium. The more helium is created in protostars the better chance they have to develop into healthy stars. The protostars that fail to develop enough helium remain dwarf stars.

Pure Energy found in all matter *(including humans)* never dies; it is always reused or recycled throughout the cosmos.

Quadrillion *(quadri + million, numerical number = 1000^5)* is written in this form: 1,000,000,000,000,000. One million multiplied by a billion will give you a quadrillion. In Germany and Great Britain this number differs greatly, which I fail to understand.

Quarks are subatomic particles that make up protons and neutrons, needed for the creation of all elements *(2 up quarks + 1 down quark + gluon = 1 proton; 1 up quark + 2 down quarks + gluon = 1 neutron)*. I firmly believe that quarks are made up of smaller micro-particles, which will be discovered in the near future.

Quintillion *(quinti + million, numerical number = 1000^6)* is written in this form: 1,000,000,000,000,000,000. One trillion multiplied by a trillion will give you a quintillion. Again, in Germany and Great Britain this number differs greatly. I ask "Why?".

Renaissance *(Il Rinascimento)* – The Renaissance began in Italy in the mid-1200s bringing Italy out of the Middle Ages into an era of enlightenment. I firmly believe that a rich Mediterranean diet positively affected and nourished the human brain and body. Literature, art, music, poetry, science and architecture flourished in Florence, Rome and in other Italian cities during the 1300s and 1400s through the works of Dante, Giotto, Giovanni Di Paolo, Fra Filippo Lippi, Leonardo Da Vinci, Michelangelo and so many other artists and authors. During this time the human brain greatly developed after greatly suffering from the Dark Age and Middle Ages. By the 1500s the Renaissance had spread throughout France, England and Germany and by the 1700s it had spread throughout the entire Europe.

Reptilian Complex is the very inner part of the human brain which controls our very survival as animals. It was the first brain to develop in reptiles and it is the first neurological development in the first trimester of pregnancy. The reptilian complex that develops in embryos by the end of the third month of pregnancy looks very much like the brain of today's reptiles. It is a primitive component of the human brain but it is also the guardian in us that warns us of possible danger.

Right Hemisphere of the human brain is the inventive, artistic and scientific part of the human brain. Anything that humans have created originates as a holistic thought or idea in our right hemisphere. It sends its creative thoughts through the *corpus callosum*, a bridge of a few hundred thousands neurological fibers, to the left hemisphere where it takes the idea and breaks down into logical steps to possibly arrive towards the development of that creative thought.

Saint Joseph Teachers' College was a teachers' college in Montreal, Canada. It was incorporated with McGill University in the Education Department in the 1970s.

Santa Cruz is the capital of Santa Cruz de Tenerife in the beautiful tropical Canary Islands, belonging to Spain, off the coast of Morocco.

Schemata is an outline or diagram of the important factors which result in explanation of the formation of all the pertinent categories of a phenomenon or of a cosmic development.

Serotonin *($C_{10}H_{12}N_2O$) is* an important neurotransmitter that works in the central nervous system, in blood platelets and in our stomach and intestines. It helps us to feel good and be happy; it also helps our memory. Serotonin helps us as we age: the more serotonin present the better is our memory as we age. Omega-3 oils and the vitamin-Bs are important in preserving the presence of serotonin; low presence of serotonin can lead to memory loss, depression, violence and suicide.

Sextillion *(sext + million, numerical number = 1000^7)* is written in this form: 1,000,000,000,000,000,000,000. One trillion multiplied by a quadrillion will give you a sextillion. Again, in Germany and Great Britain this number differs greatly!

The **Solar System** *(contains about 14-15 planets and over 100 moons)* travels at approximately 864,000 km/hr. around the Milky Way.

Space-time or **spacetime** is a combination of location and the point of occurrence in our four-dimensional universe. Each space-time is unique throughout the cosmos because it depends on its location and its gravitational attraction to a cosmic object. Space-time on Earth is found only on this planet. Space and time work together in our lives; but, they differ outside Earth, differentiated by each unique gravitational force of each and every cosmic object.

Stanford-Binet Intelligence Scale is an intelligence test administered to each individual child to diagnose his or her intellectual level. It was created by Alfred Binet, a French psychologist, in 1905, and it was revised by Lewis Terman, a psychologist at Stanford University, in 1916. Although today's Stanford-Binet Intelligence Scale has gone through five editions, its top score still remains at 160 I.Q.

Subatomic particles are the smaller components that make up protons and neutrons and even smaller ones to fill the gaps inside the nuclei of all elements. There are six types of quarks, electrons, three known types of neutrinos, three types of bosons and gluon *(used to join the quarks)*. I am including **alexion** *(a new theoretical substance that I have established to fortify my theory)* which protects all subatomic particles inside all nuclei, even inside the destructive forces of black holes and in all galactic nebulae.

The **Sun** is our beautiful star that is extremely healthy by fusing 27% of its hydrogen into approximately 25% helium and 2% carbon, nitrogen, oxygen, iron and all the other elements needed to create life on planet earth. The sun's surface reaches approximately 5700°C and its core averages 15,000,000°C. Our sun produces solar winds that constantly hit our planet up to 400 kilometers per second: that is approximately 1,440,000 kilometers per hour. Periodical gigantic solar flares create hot temperatures on earth and can be destructive by causing magnetic storms that damage our electrical power stations. Presently it takes the sun approximately 205 million years to revolve around the Milky Way at approximately 27,000 light years from its center in the Orion arm. During the creation of the sun in one of the suburban nebula its position was approximately 50,000-55,000 light years away from the center of the Milky Way and its first revolution might have taken 500-600 million years.

Supreme Being, Supreme Matter, and **Supreme Power** are the names I use for God, who is forever present in every subatomic particle in the cosmos and works within every gravitational center of every cosmic body *(especially within the destructive forces of each and every black hole)*.

Synapses are microscopic gaps between the dendrites of the axons of neurons that facilitate the continuous travel of neurotransmitters throughout approximately 260,000 miles of neurological network. There are more than a trillion synapses that connect hundreds of billions neurons and glial cells, permitting quick messaging and retrieval of memory and many other important functions of all neurotransmitters.

Temporal Lobe processes sound and smell that enter the brain and spoken language by interpreting the sounds of words.

Tertiary Period of the Cenozoic Era *(approximately 3,800,000 years ago)* was the beginning of the development of the neocortex in hominids, wrapped around the mammalian brain and reptilian complex and neurologically connected to both of them.

Teotihuacán *(means The City of the Gods)* is an ancient Meso-American city about 41 kilometers northeast of Mexico City *(The dirt road up the mountains from outside Mexico City to Teotihuacán is dangerous and treacherous. Only one vehicle can fit, creating serious problems if there is another vehicle coming down. Also, heavy rains cause parts of the dirt road to break, making it very difficult to go on. Hopefully, by now it has been fixed and improved.)* It is famous for the beautiful Pyramids of the Sun, of the Moon, and of Quetzalcoatl. It was well planned with beautiful long boulevards and streets. When Rome was the largest city on earth during the First century *(more than a million citizens)* Teotihuacán was the second largest city in the Americas with more than 200,000 people. It was definitely built by the Toltecs during the first millennium B.C. Some archaeologists believe that it might have been the Totonacs *(who built the beautiful and enchanting El Tajin)*. However, after visiting and studying both ancient cities I disagree that the Totonacs might have built Teotihuacán because the architecture of El Tajin is much different from it.

Thalamus is in the center of our brain; it interprets all information from our senses, except smell, and sends that information to the appropriate parts of the neocortex.

Tikal *(originally called Yax Mutal in Mayan)* was the largest Mayan city built from the Fourth Century B.C. on in the Guatemalan jungle, about 300 kilometers northeast of Guatemala City. It reached its Classic Period around 200 A.D. and it was conquered by Teotihuacán in the 300s A.D. This gorgeous ancient city is filled with wondrous steep pyramids, temples, palaces and a beautiful acropolis: it is called the New York City of the Mayas. At its height, Tikal had a population of nearly half a million people. You can hire a small plane from Guatemala City to get to a small landing strip near Tikal. I decided to drive from Cancun to Chetumal, to Belize City, to Belmopan, to San Ignacio and cross the border at Benque Viejo and continue on a dirt road to Tikal. I have visited more than 100 ancient cities around the world and Tikal is one of my favorites. If you decide to drive on the route I took, make sure you do so on dry season. The rainy season can be treacherous, turning dirt roads into rivers. Plan to stay at Tikal at least for a week: there are so many steep pyramids to climb and so many beautiful temples and palaces to visit, especially its acropolis.

Triassic Period *(approximately 250-200 million years ago)* is when the first mammals developed, as the mammalian brain began to develop around the reptilian complex, neurologically connected to it.

Triune Brain was conceived by the great neuroscientist Paul MacLean when he discovered that the human brain is made up of three distinctive, yet neurologically connected sub-brains: the Reptilian complex, the Mammalian Brain and the Neocortex.

WISC – Wechsler Intelligence Scale for Children is an I.Q. test given to children between 6-16 years old. It was developed by David Wexler in 1939; and, he reorganized it in 1949. It is made up of 10 subtests; it has been revised five times in the past seventy years.

Yin & Yang is an ancient symbol *(darkness + light, negative + positive)* that realistically describes how cosmic bodies work together. According to ancient Chinese philosophy everything in the universe contains yin & yang. Yin & yang work together; they do not oppose each other. In the cosmos matter *(positive)* revolves around gravity *(negative)*. Together, cosmic bodies work harmoniously around their respective gravitational forces. Everything in the cosmos is ruled by yin & yang *(negative & positive forces)*. In the atomic and subatomic level the nuclei of elements are positively charged *(protons)* and the electrons are negatively charged, working perfectly together.

BIBLIOGRAPHY

(Related books, articles and presentations)

Aamodt, Sandra and Sam Wang (2008). *Welcome to Your Brain.* New York, N.Y.: MJF Books.
Abraham, Carolyn (2001). *Possessing Genius: the bizarre odyssey of Einstein's brain.* Toronto, Canada: Penguin Canada.
Abrahams, Peter (2015, editor). *How the Brain Works.* New York, N.Y.: Metro Books.
Aczel, Amir (1996). *Fermat's Last Theorem.* New York, N.Y.: Delta.
Aczel, Amir D. (1999). *God's Equation: Einstein, Relativity, and the Expanding Universe.* New York, N.Y.: Delta.
Aguilar, David A. (2007). *Planets, Stars, and Galaxies.* Washington, D.C.: National Geographic Society.
Amen, Daniel G. (2005). *Making a Good Brain Great.* New York, N.Y.: Three Rivers Press.
Amen, Daniel G. (2008). *Magnificent Mind at Any Age.* New York, N.Y.: Harmony Books.
Andreasen, Nancy C. (2001). *Brave New Brain: conquering mental illness in the era of the genome.* New York, N.Y.: Oxford University Press.
Barnett, Lincoln (1948). *The Universe and Dr. Einstein.* New York, N.Y.: Bantam Books.
Bartusiak, Marcia (2015). *Black Hole: how an idea abandoned by Newtonians, hated by Einstein, and gambled on by Hawking became loved.* New Haven, MA.: Yale University Press.

Baumann, Mary K., Will Hopkins, Loralee Nolletti and Michael Soluri (2005). ***What's Out There.*** London, England: Duncan Baird Publishers Ltd.

Begelman, Mitchell and Martin Rees (1996). ***Gravity's Fatal Attraction: black holes in the universe.*** New York, N.Y.: Scientific American Library.

Bekenstein, J.D. (1980). **Black Holes Thermodynamics** in *Physics Today.* January 24.

Bennett, Jeffrey, Megan Donahue, Nicholas Schneider and Mark Voit (1999). ***The Cosmic Perspective.*** San Francisco, CA. Addison Wesley.

Bennett, Jeffrey, Seth Shostak and Bruce Jakosky (2003). ***Life in the Universe.*** San Francisco, CA. Addison Wesley.

Bennett, Jeffrey, Megan Donahue, Nicholas Schneider and Mark Voit (2007). ***The Essential Cosmic Perspective.*** San Francisco, CA. Pearson, Addison-Wesley, fourth edition.

Black, Ira B. (2002). ***The Changing Brain.*** New York, N.Y.: Oxford University Press.

Bloom, Howard (2000). ***Global Brain: the evolution of mass mind from the big bang to the 21st century.*** New York, N.Y. John Wiley & Sons, Inc.

Born, Max (1965). ***Einstein's Theory of Relativity.*** New York, N.Y.: Dover.

Brizendine, Louann (2006). ***The Female Brain.*** New York, N.Y.: Morgan Road Books.

Brown, J.L. and E. Pollitt (1996). ***Malnutrition, Poverty and Intellectual Development.*** In *Scientific American,* February, pp. 38-43.

Caplan, T. and F. Caplan (1982). ***The Second Twelve Months of Life.*** New York, N.Y.: Bantam Books.

Caplan, T. and F. Caplan (1984). ***The Early Childhood Years: the Two to Six Years Old.*** New York, N.Y.: Bantam Books.

Caplan, T. and F. Caplan (1995). ***The FirstTwelve Months of Life.*** New York, N.Y.: Bantam Books, revised edition.

Carper, Jean (2000). ***Your Miracle Brain.*** New York, N.Y.: Harper Collins Publishers.

Chandrasekhar, S. (1935). **The Maximum Mass of Ideal White Dwarfs** in *Astrophysical Journal, 74,* 81.

Chandrasekhar, S. (1957). *Eddington: The Most Distinguished Astrophysicist of His Time.* Cambridge, England: Cambridge University Press.

Chandrasekhar, S. *The Black Hole in Astrophysics: The Origin of the Concept and Its Role.* Contemporary Physics 15 (1974): 1-24.

Chandrasekhar, S. (1992). *The Mathematical Theory of Black Holes.* New York, N.Y.: Oxford University Press, Inc.

Comins, Neil F. and William J. Kaufmann III (2008). *Discovering the Universe.* New York, N.Y.: W.H. Freeman and Company.

Conway, John and Richard Guy (1996). *The Book of Numbers.* New York, N.Y.: Copernicus.

Cornell, James, editor (1989). *Bubbles, Voids and Bumps in Time: the new cosmology.* Cambridge, England: Cambridge University Press.

Croswell, Ken (1999). *Magnificent Universe.* New York, N.Y.: Simon & Schuster.

Davidson, J. and B. Davidson (2004). *Genius Denied: How to Stop Wasting Our Brightest Young Minds.* New York, N.Y.: Simon & Schuster.

Davies, Paul (1983). *God and the New Physics.* New York, N.Y.: Viking Penguin.

Davies, Paul (1991). *The Mind of God.* New York, N.Y.: Simon & Schuster.

Davies, Paul (1995). *About Time: Einstein's unfinished revolution.* New York, N.Y. Simon & Schuster.

Denis, Brian (1997). *Albert Einstein.* New York, N.Y.: Wiley.

De Duve, Christian (1995). *Vital Dust: Life as a cosmic imperative.* New York, N.Y.: Basic Books.

DeGrasse Tyson, Neil (2007). *Death by Black Hole and Other Cosmic Quandaries.* New York, N.Y.: W.W. Northorn & Company.

Di Paolo, Vincent L. (September 1, 1993). *Synchronizing Teaching with Neurological development.* Manassas, Virginia: The Educator as an Instrument of Change: Professional Development Conference, Prince William County Public Schools, 90 minutes.

Di Paolo, Vincent L. (November 5, 1993). ***Synchronizing the Teaching of Speaking, Reading, and Writing With Children's Brain Development: a brain development learning theory.*** Virginia Beach, Virginia: The ESL Statewide Training Institute, 70 minutes.

Di Paolo, Vincent L. (October 14, 1994). ***Synchronizing the Curriculum with the Neurological Development of Children.*** Woodbridge, Virginia: Phi Delta Kappa, 2 hours.

Di Paolo, Vincent L. (August 30, 1995). ***Neurological Development of Middle school Children: implications for instructions.*** Woodbridge, Virginia: Lake Ridge M.S., Prince William County Public Schools, two presentations of 150 minutes each: 9:00-11:30 a.m. and 1:00-3:30 p.m.

Di Paolo, Vincent L. (September 29 and November 10, 1995). ***Implications for Instruction to Right Brain Students with ADD/ADHD.*** Fork Union, Virginia Fork Union Military Academy, two 3-hour presentations.

Di Paolo, Vincent L. (March 2, 1996). ***The Brain and Education: synchronizing the curriculum with the neurological development of children.*** Williamsburg, Virginia: 1996 Virginia ESL Conference, 2 hour presentation.

Di Paolo, Vincent L. (August 26- 27, 1996). ***The Brain and Education: synchronizing the curriculum with the neurological development of children.*** Manassas, Virginia: Building Bridges to Lifelong Learning Conference, Prince William County Schools, Two 150-minute presentations.

Di Paolo, Vincent L. (March 7, 1997). ***The Brain and Education: the triune brain and the four myelinations.*** Norfolk, Virginia: 1997 Virginia Middle School Association Annual Conference, 2 hour presentation.

Di Paolo, Vincent L. (April 18, 1997). ***The Brain, Education and the Acquisition of Second Languages.*** McLean, Virginia: Facing the Daily Challenges of Your Multilingual Classroom Conference, 2 hour presentation.

Di Paolo, Vincent L. (May 8, 1997). *The Brain, Education and the Acquisition of Second Languages.* Charlottesville, Virginia: Piedmont ESL Roundtable Spring Workshop, 2 hour presentation.

Di Paolo, Vincent L. (February 4, 1999). *The Brain and Education: the acquisition of reading and children at risk.* Washington, D.C.: Oral presentation to the United States Department of Education, Institute for Professional Development, Office of Elementary and Secondary Education, 90 min. presentation.

Di Paolo, Vincent L. (March 12, 1999). *The Brain, Education and Memory.* Norfolk, Virginia: 1999 Virginia Middle School Association Annual Conference, 70 minute presentation.

Di Paolo, Vincent L. (November 2, 1999). *The Brain, Education and Children At-Risk: how can we help them to learn more, score higher, and succeed in life?* Manassas, Virginia: Oral presentation at Osbourn H.S., Prince William County Public Schools Student Services Symposium, 2 hour presentation.

Di Paolo, Vincent L. (August 18, 2003). *Synchronizing the Curriculum with the Stages of Brain Growth and How Nutrition Affects Learning, Memory, and Test-Taking.* Manassas, Virginia: Oral presentation at the Pennington School, Prince William County Schools, 3 hours a.m. and 3 hours p.m.

Di Paolo, Vincent L. (March 12, 2004). *The Omega-3 State: how nutrition affects learning, memory, and health!* Norfolk, Virginia: 2004 VMSA Annual Conference, 2 hour presentation.

Di Paolo, Vincent L. (March 27, 2004). *The Omega-3 State: how nutrition affects learning, memory, and health!"* Bristow, Virginia: 2004 "Girls + Math + Science = SUCCESS!" Conference, Marsteller M.S., Prince William County Public Schools, 2 hour presentation.

Di Paolo, Vincent L. (2004). *My Beloved Friend, Judas.* Bloomington, IN.: AuthorHouse.

Di Paolo, Vincent L. (March 11, 2005). *Omega-3, the State of Learning.* Norfolk, Virginia: 2005 VMSA Annual Conference, 90 minute presentation.

Di Paolo, Vincent L. (November 3, 2005). *The Brain and Education: how nutrition affects neurological growths, learning, memory, attention, and health.* Philadelphia, PA.: National Middle School Association, 75 minute presentation.

Di Paolo, Vincent L. (March 25, 2006). ***Excellent Nutrition = Excellent Brain & Body.*** Bristow, Virginia: 2006 "Girls + Math + Science = SUCCESS!" Conference, Marsteller M.S., Prince William County Public Schools, 2 hour presentation.

Di Paolo, Vincent L. (March 29, 2008). ***Girls and Boys: gender curricula.*** Bristow, Virginia: 2008 "Girls + Math + Science = SUCCESS!" Conference, Marsteller M.S., Prince William County Public Schools, 2 hour presentation.

Di Paolo, Vincent L. (March 28, 2009). ***Myelination of the Right Hemisphere During Middle School Years: how the arts affect science, Math, and languages in the left hemisphere.*** Bristow, Virginia: 2009 "Girls + Math + Science = SUCCESS!" Conference, Marsteller M.S., Prince William County Public Schools, 2 hour presentation.

Di Paolo, Vincent L. (March 20, 2010). ***Right Brain, Left Brain: learning and memory.*** Bristow, Virginia: 2010 "Girls + Math + Science = SUCCESS!" Conference, Marsteller M.S., Prince William County Public Schools, 2 hour presentation.

Di Paolo, Vincent L. (March 26, 2011). ***Fine Arts & Nutrition and Myelination in the Right Hemisphere.*** Bristow, VA.: 2011 "Girls + Math + Science = SUCCESS!" Conference, Marsteller M.S., Prince William County Public Schools, 2 hour presentation.

Di Paolo, Vincent L. (March 24, 2012). ***The Fine Arts, Nutrition & the Expansion of the Right Hemisphere: how the arts affect memory and the learning of math and science.*** Bristow, Virginia: 2012 "Girls + Math + Science = SUCCESS!" Conference, Prince William County Public Schools, 2 hour presentation.

Di Paolo, Vincent L. (2015). ***My Beloved Friend, Judas.*** Bloomington, IN.: XLibris, second edition with new format.

DoCarmo, Manfredo (1976). ***Differential Geometry of Curves and Surfaces.*** Englewood Cliffs, N.J.: Prentice-Hall.

Eddington, Sir Arthur S. (1923). ***The Mathematical Theory of Relativity.*** New York, N.Y.: Cambridge University Press.

Eddington, Sir Arthur S. (1959). *Space, Time and Gravitation: An Outline of the General Relativity Theory.* New York, N.Y.: Harper & Row.

Educational Leadership (1998). *How the Brain Learns.* ASCD, Nov., 1998, Vol.56, No.3.

Einstein, Albert and Leopold Infeld (1938). *The Evolution of Physics: from early concepts to relativity and quanta.* New York, N.Y.: A Clarion Book Published by Simon and Schuster, ninth paperback printing.

Einstein, Albert (1950). *The Theory of Relativity & Other Essays.* New York, N.Y.: MJF Books.

Einstein, Albert (1950). *Out of My Later Years.* New York, N.Y.: Philosophical Library.

Einstein, Albert (1952). *Relativity: the Special and the General Theory.* New York, N.Y.: Crown Publishers, Inc., 1961.

Einstein, Albert (1956). *Out of My Later Years.* Avenel, N.J.: Wings Books, 1993.

Einstein, Albert. *The Philosophy of Albert Einstein.* Edited by Walt Martin and Magda Ott. New York, N.Y. Fall River Press, 2013.

Eisenbud, Leonard (1971). *The Conceptual Foundations of Quantum Mechanics.* New York, N.Y.: Van Nostrand Reinhold.

Eliot, Lise (1999). *What's Going On in There?: How the brain and mind develop in the first five years of life.* New York, N.Y.: Bantam Books.

Epstein, Herman T. **Growth spurts during brain development: Implications for Educational Policy and Practice,** *in Education and the Brain,* ed. J.S. Chall (Chicago: *National Society for the Study of Education, 1978), ch.10.*

Epstein, Herman T. **Learning to learn: Matching instruction to cognitive levels, Principal,** May 1981, pp. 25-30.

Ferguson, Kitty (1991). *Stephen Hawking: quest for a theory of everything.* New York, N.Y. Bantam Books.

Ferris, Timothy (2000). *Life Beyond Earth.* New York, N.Y.: Simon & Schuster.

Filkin, David (1997). *Stephen Hawking's Universe: the Cosmos Explained.* New York, N.Y.: Basic Books.

Fölsing, Albrecht (1997). *Albert Einstein.* New York, N.Y.: Penguin.
Frank, Bill (2003). *Forever Young: 100 age-erasing techniques.* New York, N.Y.: Harper Collins.
Frank, Phillipp (1957). *Einstein: His Life and Times.* New York, N.Y.: Knopf.
Fritzsch, Harald (1983). *Quarks.* New York, N.Y.: Basic Books.
Fuster, J. (1997). *The Prefrontal Cortex.* Philadelphia, PA.: Lippencott Raven.
Gardner, H. (1983). *Frames of Mind: The Theory of Multiple Intelligences.* New York, N.Y.: Basic Books.
Gardner, H. (1985). *The Mind's New Science: A History of the Cognitive Revolution.* New York, N.Y.: Basic Books.
Gardner, H. (1991). *The Unschooled Mind: How Schools Should Teach.* New York, N.Y.: Basic Books.
Gardner, H. (1993). *Creating Minds.* New York, N.Y.: Basic Books.
Garlick, Mark A. and Wil Tirion (2006). *The Illustrated Atlas of the Universe.* Sydney, Australia: Weldon Owen Inc.
Gazzaniga, Michael S., editor (2000). *The New Cognitive Neuroscience.* Cambridge, Mass.: MIT Press, second edition.
Geary, D.C. (1998). *Male, Female: The Evolution of Human Sex Differences.* Washington, D. C.: American Psychological Association.
Giuffre, Kenneth and Theresa Foy Di Geronimo (1999). *The Care and Feeding of your Brain.* Franklin Lakes, N.J.: Career Press.
Glendenning, Norman K. (2004). *After the Beginning: a cosmic journey through space and time.* London, England: Imperial College Press.
Goldsmith, Donald and Nathan Cohen (1991). *Mysteries of the Milky Way.* Chicago, Illinois: Contemporary Books.
Gogtay, N. et al. (2004). *Dynamic Mapping of Human Cortical Development During Childhood Through Early Adulthood.* Proceedings of the National Academy of Sciences, May, 25, 101(21), 8174-79.
Greenfield, Susan A. (1997). *The Human Brain.* New York, N.Y.: Basic Books.
Gribbin, John (1979). *Timewarps.* New York, N.Y.: Delacorte.

Gribbin, John (1984). *In Search of Schrödinger's Cat.* New York, N.Y.: Bantam Books.
Gribbin, John (1986). *In Search of the Big Bang.* New York, N.Y.: Bantam Books.
Gribbin, John (1991). *Blinded by the Light.* New York, N.Y.: Harmony Books.
Gribbin, John (1993). *In the Beginning: the birth of the living universe.* New York, N.Y.: Little, Brown and Company.
Guggenheimer, H.W. (1977). *Differential Geometry.* New York, N.Y.: dover.
Halpern, D.F. (2000). *Sex Differences in Cognitive Abilities.* Mahwah, N.J.: Lawrence Erlbaum, third edition.
Hart, Leslie A. (1983). *Human Brain and Human Learning.* Oak Creek, AR.: Books for Educators.
Hawking, S.W. (1974) *Black Hole Explosions? Nature 248*, pp. 30-31.
Hawking, Stephen (1988, 1996). *A Brief History of Time.* Tenth Anniversary Edition, New York, N.Y.: Bantam Books, 1998.
Hawking, Stephen (1991). *Quest for a Theory of Everything.* New York, N.Y.: Bantam Books.
Hawking, Stephen (1993). *Black Holes and Baby Universes and Other Essays.* New York, N.Y.: Bantam Books.
Hawking, Stephen (2001). *The Universe in a Nutshell.* New York, N.Y.: Bantam Books.
Hawking, Stephen with Leonard Mlodinow (2005). *A briefer History of Time.* New York, N.Y.: Bantam Dell.
Hawking, Stephen (2014). *The Illustrated A Brief History of Time— The Universe in a Nutshell.* New York, N.Y.: Bantam Books.
Herber, H. (1978). *Teaching Reading in Content Areas.* Englewood Cliffs, N.J. Prentice Hall, second edition.
Hey, Tony and Patrick Walters (2003). *The New Quantum Universe.* Cambridge, United Kingdom: Cambridge University Press.
Howard, Pierce J. (2006). *The Owner's Manual for the Brain: everyday applications from Mind-Brain Research.* Austin, Texas: Bard Press.

Hughes, Virginia (2013). ***Rewiring the Brain: how neuroscience will fight five age-old afflictions,*** in *Popular Science,* March 2013, pp. 34-39.

Huggett, S.A. et al. (1998). ***The Geometric Universe: Science, Geometry, and the Work of Roger Penrose.*** New York, N.Y.: Oxford University Press.

Hunt, Morton (1982). ***The Universe Within: A new science explores the human mind.*** New York, N.Y.: Simon & Schuster.

Ikenberry, Ernest (1962). ***Quantum Mechanics.*** London, England: Oxford University Press, Oxford.

Jastrow, Robert (1979). ***Red Giants and White Dwarfs.*** New York, N.Y.: W.W. Norton & Company, Inc., new edition.

Jastrow, Robert (1989). ***Journey to the Stars: space exploration-tomorrow and beyond.*** New York, N.Y.: Bantam Books.

Kaufmann, William J. III (1979). ***Black Holes and Warped Spacetime.*** San Francisco, CA.: W.H. Freeman and Company.

Kaufmann, William J. III (1998). ***Universe.*** San Francisco, CA.: W.H. Freeman & Company.

Kitchin, Chris (2007). ***Galaxies in Turmoil: the active and starburst galaxies and the black holes that drive them.*** London, England: Springer-Verlag London Limited.

Kotulak, Ron (1996). ***Inside the Brain: Revolutionary discoveries of how the mind works.*** Kansas City, MI.: Andrews and McMeel, a Universal Press Syndicate Co.

Krauss, Lawrence M. (2012). ***A Universe from Nothing.*** New York, N.Y.: Free Press.

Leach, P. (1997). ***Your Baby and Child: From Birth to Age Five.*** New York, N.Y.: Knopf.

LeDoux, Joseph (2002). ***Synaptic Self: how our brains become who we are.*** New York, N.Y.: Penguin Books.

Levy, David H. (1998). ***The Ultimate Universe: the most up-to-date guide to the cosmos.*** New York, N.Y.: Pocket Books.

Longair, Malcolm S. (1996). ***Our Evolving Universe.*** London, England: Cambridge University Press.

Luminet, J.P. (1992). ***Black Holes.*** London, England: Cambridge University Press, Cambridge.

MacLeaN, Paul D. (1973). *A Triune Concept of the Brain and Behavior.* Toronto, Ontario: University of Toronto Press.

MacLean, Paul D. (1990). *The Triune Brain: Role in paleocerebral functions.* New York, N.Y.: Plenum Press.

Marsa, Linda (2017). **What Once Was Lost: How neural stem cells repair damage from strokes, spinal injuries and aging** *in Discovery.* October, 2017, pp. 33-39.

McKhann, Guy & Marilyn Albert (2002). *Keep Your Brain Young.* New York, N.Y.: John Wiley & Sons.

McSweeney, Harry Y. Jr. (1994). *Stardust to Planets.* New York, N.Y.: St. Marin's.

Mellor, H. (1981). *Real Time.* New York, N.Y.: Cambridge University Press.

Moore, Keith L. and Arthur F. Dalley (1999). *Clinically Oriented Anatomy.* Baltimore, Maryland: Lippincott Williams & Wilkins.

Moore, Sir Patrick (2003). *Atlas of the Universe.* Toronto, Ontario, Canada: Firefly Books Limited.

Nathan, Peter (1988). *The Nervous System.* Oxford, England: Oxford University Press.

Newsweek (June 3, 1991). *Heavens! Black Holes, Quasars, Starquakes: astronomers launch a new age of discovery.* New York, N.Y.: Newsweek.

Newsweek (March 27, 1995). *The New Science of the Brain.* New York, N.Y.: Newsweek.

Nolte, David D. (2001). *Mind at Light Speed: a new kind of intelligence.* New York, N.Y.: The Free Press.

Novikov, I.D. (1990). *Black Holes and the Universe.* Cambridge, United Kingdom: Cambridge University Press.

Pais, Abraham (1982). *'Subtle Is the Lord...': The Science and the Life of Albert Einstein.* New York, N.Y.: Oxford University Press.

Pais, Abraham (1994). *Einstein Lived Here.* New York, N.Y.: Oxford University Press.

Peebles, P.J.E. (1993). *Principles of Physical Cosmology.* Princeton, N.J.: Princeton University Press.

Penrose, Roger (1989). *The Emperor's New Mind.* New York, N.Y.: Oxford University Press.

Piaget, Jean (1953). *The Origin of Intelligence in the Child.* New York, N.Y.: Routledge and Kegan Paul.

Pickover, Clifford A. (1998). *Time: a traveler's guide.* New York, N.Y.: Oxford University Press.

Primack, Joel R. and Nancy Ellen Abrams (2006). *The View from the Center of the Universe: discovering our extraordinary place in the cosmos.* New York, N.Y.: Riverhead Books.

Ramachandran, V.S. (2011). *The Tell-Tale Brain: a neuroscientist's quest for what makes us human.* New York, N.Y.: W.W. Norton & Company.

Ratey, John J. (2001). *A User's Guide to the Brain: Perception, attention, and the four theaters of the brain.* New York, N.Y.: Pantheon Books.

Rathus, Spencer A. (2008). *Childhood and Adolescence: Voyages in development.* Belmont, CA.: Thomson Learning, Inc., third edition.

Raymo, Chet ((2001). *An Intimate Look at the Night Sky.* New York, N.Y.: Walker & Company.

Reader's Digest (1980). *Foods that Harm Foods that Heal: An A-Z guide to safe and healthy eating.* Pleasantville, N.Y.: The reader's Digest Association, Inc.

Reichenbach, H. (1958). *The Philosophy of Space and Time.* New York, N.Y.: Dover.

Restak, Richard (1980). *The Brain: the last frontier.* New York, N.Y.: Warner.

Restak, Richard (1988). *The Mind.* New York, N.Y.: Bantam Books.

Restak, Richard M. (1994). *The Modular Brain: how new discoveries in neuroscience are answering age-old questions about memory, free will, consciousness, and personal identity.* New York, N.Y.: Touchstone, Simon & Schuster.

Restak, Richard (2000). *Mysteries of the Mind.* Washington, D.C.: National Geographic Society.

Restak, Richard (2003). *The New Brain.* U.S.A.: Rodale, Inc.

Restak, Richard (2006). *The Naked Brain: how the emerging neurosociety is changing how we live, work, and love.* New York, N.Y.: Harmony Books.

Ridpath, Ian (1991). *Astronomy: how we view our solar system and the universe beyond.* New York, N.Y.: Gallery Books.

Robbins, John (1987). *Diet for a New America: How your food affects your health, happiness, and the future of life on earth.* Tiburon, CA.: H.J. Kramer Book.

Rolfs, Claus E. and William S. Rodney (1988). *Cauldrons in the Cosmos: nuclear astrophysics.* Chicago, Illinois: The University of Chicago Press.

Rose, Steven (2005). *The Future of the Brain: the promise and perils of tomorrow's neuroscience.* New York, N.Y.: Oxford University Press.

Ryan, William and Walter Pitman (1998). *Noah's Flood: the new scientific discoveries about the event that changed history.* New York, N.Y.: Touchstone.

Sagan, Carl (1977). *The Dragons of Eden: speculations on the evolution of human intelligence.* New York, N.Y.: Random House.

Sagan, Carl (1980). *Cosmos.* New York, N.Y.: Random House.

Sagan, Carl (1994). *Pale Blue Dot: a vision of the human future in space.* New York, N.Y.: Random House.

Sagan, Carl and Ann Druyan (1992). *Shadows of Forgotten Ancestors.* New York, N.Y.: Ballantine Books.

Santrock, Jofn W. (2009). *Child Development.* New York, N.Y.: McGraw Hill, twelfth edition.

Sayen, J. (1985). *Einstein in America.* New York, N.Y.: Crown.

Scalzi, John (2003). *The Rough Guide to the Universe.* London, England: Rough Guides Ltd.

Scharf, Caleb (2012). *Gravity's Engines: how bubble-blowing black holes rule galaxies, stars, and life in the cosmos.* New York, N.Y.: Scientific American/Farrar, Straus and Giroux.

Schatzman, E. (1992). *Our Expanding Universe.* New York, N.Y.: McGraw-Hill.

Schiffman, Harvey Richard (2001). *Sensation and Perception.* New York, N.Y.: John Wiley & Sons, Inc.

Scientific American (2009). *The Evolution of Evolution.* January, 2009.

Sears, Barry (2003). *The Omega Rx Zone: the miracle of the new high-dose fish oil.* New York, N.Y.: Regan Books/Harper Collins.

Sedvall, G., L. Farde, A. Persson and FA Wiesel (1986). **Imaging of neurotransmitter receptors in the living human brain** *in Archives of General Psychiatry.* 43: pp. 995-1005.

Seeds, Michael A. (2003). *Stars and Galaxies.* Pacific Grove, CA. Thomson: Brooks/Cole.

Seife, Charles (2000). *Zero: the biography of a dangerous idea.* New York, N.Y.: Penguin Books.

Shapiro Stuart L. and Saul A. Teukolsky (1983). *Black Holes, White Dwarfs and Neutron Stars: The Physics of Compact Objects.* New York, N.Y.: John Wiley and Sons, 1st edition.

Silk, Joseph (1994). *A Short History of the Universe.* New York, N.Y.: Scientific American Library.

Silk, Joseph (2001). *The Big Bang.* New York, N.Y.: W.H. Freeman & Co. Third Edition.

Singh, Simon (1997). *Fermat's Enigma.* New York, N.Y.: Walker and Company.

Singh, Simon (2004). *Big Bang: the origin of the universe.* New York, N.Y.: Fourth Estate, an Imprint of Harper Collins Publishers.

Smart, J.J.C., editor (1964). *Problems of Space and Time.* New York, N.Y.: MacMillan.

Sparrow, Giles (2006). *Cosmos.* London, England: Quercus Publishing Ltd.

Sparrow, Giles (2007). *The Stargazer's Handbook: an atlas of the night sky.* London, England: Quercus Publishing Ltd.

Sparrow, Giles (2014). *Hubble: legacy edition.* New York, N.Y.: Metro Books.

Springer, Sally and Georg Deutsch (1989). *Left Brain, Right Brain.* New York, N.Y.: W.H. Freeman and Company.

Stein, James D. (2011). *Cosmic Numbers: the numbers that define our universe.* New York, N.Y.: Basic books.

Stoll, Andrew L. (2001). *The Omega-3 Connection.* New York, N.Y.: Fireside.

Susskind, Leonard (2008). *The Black Hole War: my battle with Stephen Hawking to make the world safe for quantum mechanics.* New York, N.Y. Little, Brown and Company.

Sweeney, Michael S. (2013). *Complete Guide to Brain Health.* Washington, D.C.: National Geographic.

Tayler, R.J. (1993). ***Galaxies: Structure and Evolution.*** Cambridge, United Kingdom: Cambridge University Press.

Taylor, E. and J. Wheeler. (1992). ***Spacetime Physics.*** New York, N.Y.: W.H. Freeman.

Teilhard de Chardin, Pierre (1959). ***The Phenomenon of Man.*** New York, N.Y.: Harper and Row.

Terzian, Yervant and Elizabeth Bilson, editors (1997). ***Carl Sagan's Universe.*** Cambridge, United Kingdom: Cambridge University Press.

Tegmark, Max (2014). ***Our Mathematical Universe: My Quest for the Ultimate Nature of Reality.*** New York: Alfred A Knopf.

Thorne, Kip S. (1994). ***Black Holes and Time Warps: Einstein's Outrageous Legacy.*** New York: W.W. Norton.

Trattler, Ross and Adrian Jones (2001). ***Better Health through Natural Healing: How to get well without drugs or surgery.*** Dingley, Victoria, Australia: Hinkler Books.

Trefil, James S. (1985). ***Space Time Infinity.*** Washington, D.C.: Smithsonian Books.

Trefil, James (1994). ***From Atoms to Quarks: an introduction to the strange world of particle physics.*** New York, N.Y.: Anchor Books.

Trefil, James (1999). ***Other Worlds: Images of the Cosmos from Earth and Space.*** Washington, D.C.: National Geographic Society.

Trefil, James (2012). ***Space Atlas: Mapping the Universe and Beyond.*** Washington, D.C.: National Geographic.

Wald, R.M. (1977). ***Space, Time and Gravity.*** Chicago, Illinois: University of Chicago Press.

Weinberg, Steven (1972). ***Gravitation and Cosmology: Principles and Applications of the General Theory of Relativity.*** New York, N.Y.: Wiley.

Weintraub, David A. (2011). ***How Old Is the Universe?*** Princeton, N.J.: Princeton University Press.

Wheelwright, Jeff (2017). **This Old Brain** in ***Discover.*** October, 2017, pp. 27-31.

Wittrock, M.C. (1977). ***The Human Brain.*** Englewood Cliffs, N.J.: Prentice Hall.

Wolfe, H.E. (1945). ***Non-Euclidean Geometry.*** New York, N.Y.: Holt, Rinehart and Winston.

Yates, J.C. (1990). ***The Timelessness of God.*** Lanham, MD.: University Press of America.

Zakariya, Sally Banks. ***His brain, her brain: A conversation with Richard M. Restak, Principal,*** May 1981, pp. 46-51.

CREDITS FOR PHOTOGRAPHS, PAINTINGS, DRAWINGS AND CHARTS

Cover: Painting of a galaxy during regrouping of subatomic particles (Vincent L. Di Paolo)

Figure 1: NASA-HS20: Telephotos of a section of our universe taken by the Hubble telescope.

Figure 2: NASA- 3 telephotos by the Spitzer Space telescope.

Figure 3: Painting of our universe bordering other universes (Vincent L. Di Paolo).

Figure 4: Painting of the Milky Way (Vincent L. Di Paolo).

Figure 5: Five sketches illustrating a large star developing into a stellar black hole (Vincent L. Di Paolo).

Figure 6: Painting of a galactic black hole accepting millions of old stars to be recycled (Vincent L. Di Paolo).

Figure 7: Painting of the pulverization of stars inside a galactic black hole (Vincent L. Di Paolo).

Figure 8: Painting of geysers of subatomic particles being propelled into the cosmos (Vincent L. Di Paolo).

Figure 9: Painting of a galaxy as new nebulae are being formed (Vincent L. Di Paolo).

Figure 10: Formula of the probability of sun-like stars in our universe (mathematics by Vincent L. Di Paolo).

Figure 11: Drawing of the early formation of a galactic black hole by first accretion of stellar black holes (Vincent L. Di Paolo).

Figure 12: Drawing of the maturation of a galactic black hole by accretion of stellar black holes (Vincent L. Di Paolo).

Figure 13: NASA- Telephoto of Planet earth.

Figure 14: Development of the Triune Brain during pregnancy (Drawings by Vincent L. Di Paolo).

Figure 15: Left Hemisphere of the human brain (Drawing by Vincent L. Di Paolo).

Figure 16: Myelination of an axon of a neuron in human brains (Vincent L. Di Paolo).

Figure 17: Multiplication and division table exercise (Vincent L. Di Paolo).

Figure 18: Completed multiplication and division table for verification (Vincent L. Di Paolo).

Figure 19: Chart of Perfect Squares (Mathematical exercise by Vincent L. Di Paolo).

Figure 20: Chart of Perfect Cubes (Mathematical exercise by Vincent L. Di Paolo).

ABOUT THE AUTHOR

Vincent L. Di Paolo is a well-known archaeological artist, educator, scientist and writer. He has been fascinated with the cosmos since his first Cosmology course in 1966; and, he has been studying the human brain since 1980. He studied at Saint Joseph Teacher's College, at Loyola, at McGill University, at University of Virginia in Northern Virginia and at George Mason University. He has taught physics, chemistry, algebra, geometry, world history and geography, physical education, ESOL and fine arts from 1970 to 2015.

As an artist, Vincent has produced hundreds of paintings, ink drawings, bronze sculptures, and over two thousands aquatints and lithographs since 1954. He has done over seventy art shows and his work is found in over thirty countries. Many of his shows were for children's benefits *(to raise money for hospitals and schools)*. He has traveled extensively exploring and painting ancient ruins and animals in their habitats throughout the world.

Since 1980 he has been studying and researching on how the brain develops and how it learns, on the differentiated myelinations between boys and girls, and how nutrition affects myelin production, neurotransmitters, learning, memory and longevity.

He has practiced martial arts *(Judo, Wing Chun and Tae-kwon-do)* since 1963. He is also a mountaineer and rock-climber. Presently, he is writing *"Children of Yahweh"*, the sequel to his *"My Beloved Friend, Judas"*. He is also working on the following novels: *"Angel of Light"*, *"Orion"*, *"The Secret of Sierra Madre de Dios"*, and *"Kit Carson and the Amulet of the Gods"*. He is divorced and lives in Northern Virginia with Joey and Abigail, his beautiful little *Shih-tzu* and gorgeous *Calico* cat. He has two adult children, Alexandra and David.

www.ingramcontent.com/pod-product-compliance
Lightning Source LLC
Chambersburg PA
CBHW030758180526
45163CB00003B/1074